Wireless Networks

Series Editor

Xuemin Sherman Shen, University of Waterloo, Waterloo, ON, Canada

The purpose of Springer's Wireless Networks book series is to establish the state of the art and set the course for future research and development in wireless communication networks. The scope of this series includes not only all aspects of wireless networks (including cellular networks, WiFi, sensor networks, and vehicular networks), but related areas such as cloud computing and big data. The series serves as a central source of references for wireless networks research and development. It aims to publish thorough and cohesive overviews on specific topics in wireless networks, as well as works that are larger in scope than survey articles and that contain more detailed background information. The series also provides coverage of advanced and timely topics worthy of monographs, contributed volumes, textbooks and handbooks.

** Indexing: Wireless Networks is indexed in EBSCO databases and DPLB **

More information about this series at http://www.springer.com/series/14180

Choong Seon Hong • Latif U. Khan •
Mingzhe Chen • Dawei Chen • Walid Saad •
Zhu Han

Federated Learning for Wireless Networks

Choong Seon Hong
Department of Computer Science &
Engineering
Kyung Hee University, Seoul, South Korea
Gyeonggi-do
Korea (Republic of)

Latif U. Khan
Department of Computer Science &
Engineering
Kyung Hee University, Seoul, South Korea
Gyeonggi-do
Korea (Republic of)

Mingzhe Chen
Department of Electrical Engineering
Princeton University, Princeton
United States
Princeton, NJ, USA

Dawei Chen
Department of Electrical & Computer
Engineering
University of Houston, TX, United States
Houston, TX, USA

Walid Saad
Bradely Department of Electrical &
Computer Engineering
Virginia Polytechnic Institute and State
University, Blacksburg, United States
Blacksburg, VA, USA

Zhu Han
Department of Electrical & Computer
Engineering
University of Houston, TX, United States
Houston, TX, USA

ISSN 2366-1186   ISSN 2366-1445 (electronic)
Wireless Networks
ISBN 978-981-16-4965-3   ISBN 978-981-16-4963-9 (eBook)
https://doi.org/10.1007/978-981-16-4963-9

This Springer imprint is published by the registered company Springer Nature Singapore Pte Ltd.
The registered company address is: 152 Beach Road, #21-01/04 Gateway East, Singapore 189721,
Singapore

Preface

A remarkable interest in machine learning-based schemes as key enablers for next-generation intelligent wireless systems has been observed. Most of the existing learning-based solutions rely on centralized training and inference processes. However, these machine learning paradigms based on centralized training result in end users' privacy leakage and are infeasible due to large bandwidth requirements for the transfer of the enormous amount of data. Furthermore, these schemes may violate the strict latency constraints of wireless systems. To address these issues, training in a distributed machine learning scheme at the network edge can be one of the promising solutions. Distributed machine learning avoids uploading the end-devices data to a central server for training; this not only helps preserve privacy but also reduces network traffic congestion. Federated learning (FL) is one of the most important distributed learning algorithms. In particular, FL enables devices to train a shared machine learning model while keeping data locally. However, in FL, training machine learning models requires communication between wireless devices and edge servers over wireless links. Therefore, impairments of the wireless channel, such as interference, uncertainties among wireless channel states, and noise will significantly affect the FL performance. For instance, the convergence time of FL is significantly affected by the channel transmission delay. In consequence, wireless network performance optimization is necessary for wireless FL. On the other hand, FL can also be used for solving wireless communication problems and optimizing network performance. The goal of this book is to provide a comprehensive study of federated learning for wireless networks. The book consists of three main parts: (a) Fundamentals and Background of Federated Learning for Wireless Networks, (b) Design and Analysis of Federated Learning Over Wireless Networks, and (c) Federated Learning Applications in Wireless Networks. The first part deals with a brief discussion on the fundamentals of federated learning for wireless networks. In the second part, we comprehensively discuss the design and analysis of wireless federated learning. Specifically, resource optimization, incentive mechanism, security, and privacy are considered. Moreover, we present several solutions based on optimization theory, graph theory, and game theory to optimize the performance of federated learning over wireless networks. In the final part, we present several applications of federated learning in wireless networks.

Seoul, South Korea   Choong Seon Hong
Seoul, South Korea   Latif U. Khan
Princeton, NJ, USA   Mingzhe Chen
Houston, TX, USA   Dawei Chen
Blacksburg, VA, USA   Walid Saad
Houston, TX, USA   Zhu Han
May 2021

Acknowledgement

This work was supported by the National Research Foundation of Korea(NRF) grant funded by the Korean government(MSIT) (No. No. 2020R1A4A1018607) and by the Institute of Information & Communications Technology Planning & Evaluation (IITP) grant funded by the Korea government(MSIT) (No.2019-0-01287, Evolvable Deep Learning Model Generation Platform for Edge Computing). Dr. CS Hong (email:cshong@khu.ac.kr) is the corresponding author.

Contents

Part I
Fundamentals and Background

Chapter 1
Introduction

Abstract In this introductory chapter, we provide an overview of machine learning towards enabling wireless systems. The current challenges in realizing machine learning-enabled wireless systems are also presented. Additionally, we present distributed machine learning schemes. Finally, an overview of federated learning and the organization of the book is presented.

1.1 Machine Learning for Wireless Networks

An unprecedented proliferation of the Internet of Things (IoT) devices is witnessed in almost every aspect of life. These IoT devices are generating a significant amount of data that can be used by machine learning models to make the applications smarter. According to statistics, the worldwide IP traffic will reach 3.3 zettabytes by end of 2021 [1, 2]. Moreover, the smartphone traffic will exceed the PC traffic by the same year. The current wireless networks will face higher capacity demands to handle an enormous amount of traffic. Meanwhile, smart applications (e.g., extended reality, haptics) will require extremely low latency. To meet the demands of growing capacity along with low latency for various applications, there is a need to design a new wireless system. Fifth-generation (5G) wireless networks were proposed to meet the growing demands of end-users. However, there are some Internet of Everything (IoE) applications that seem difficult to be fulfilled by 5G [3–5]. These IoE applications include flying vehicles, haptics, brain-computer interaction, and extended reality among others. The IoE applications have diverse requirements in terms of latency, reliability, and user-defined metrics [6]. Therefore, to efficiently enable these applications, there is a need to propose new wireless systems, namely, the sixth-generation (6G) wireless systems. To efficiently enable these applications, 6G will rely on complex mathematical models that are difficult to model using conventional mathematical tools. Additionally, there might be a lack of mathematical models for some of the components for wireless systems [7]. On the other hand, to optimally solve various problems (e.g., wireless resource optimization, computing resource optimization, and caching decision) by using conventional mathematical optimization, we will face many challenges and

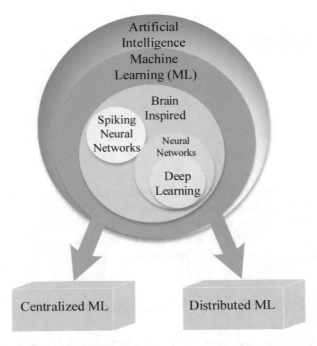

Fig. 1.1 Artificial intelligence classification

it is difficult to accurately solve these problems. To address the aforementioned challenges, one can use machine learning.

Machine learning has been applied to many applications in wireless systems. These applications include self-powered and sustainable multi-access edge computing [8], intelligent resource slicing for coexistence of enhanced Mobile Broadband and Ultra Reliable Low Latency Communications [9], intelligent caching for autonomous driving cars [10], vehicle-RSU association [11], resource management for network virtualization[12], and edge computing resource management [13], among others. An overview of artificial intelligence classification is given in Fig. 1.1. Furthermore, overview of machine learning in enabling 6G systems are shown in Fig. 1.2. Machine learning algorithms can be deployed at different layers of the network: physical layer, MAC layer, network layer, and application and transport layer. Examples of physical layer algorithms and other higher layer algorithms are link adaptation and mobility management, respectively. Mostly, machine learning-based algorithms are currently deployed statically. Moreover, these algorithms are allowed to update themselves as per the network dynamics for improved performance. From the aforementioned discussions, we can say that machine learning can be considered an integral part of the next-generation wireless systems.

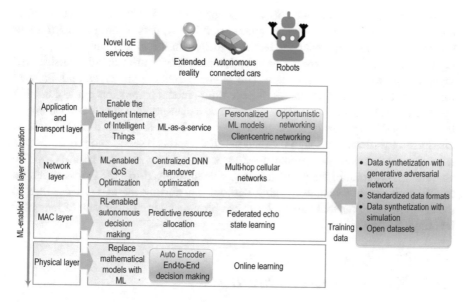

Fig. 1.2 Role of machine learning in 6G [7]

1.1.1 Current Challenges

Most of the previous works on machine learning-based wireless networks used centralized machine learning [10, 12]. The centralized machine learning is based on the migration of end-devices data to a centralized server for training. There are two main issues associated with the centralized machine learning: long training time for large datasets and privacy leakage.

- **C-1:** Training a centralized machine learning model for very large datasets (e.g., astronomical data [14]) have long training time. Additionally, it might be difficult to provide large computing power required for training of models for large datasets within a certain amount of time.
- **C-2:** The other issue that is faced by centralized machine learning-enabled wireless systems is privacy leakage due migrating the devices data to the centralized server for training [15, 16]. End-devices privacy will be leaked in case of a malicious centralized server or security attack.

1.1.2 Distributed Machine Learning

To solve the aforementioned challenge (**C-1**), one can use distributed machine learning. Distributed machine learning performs learning at various geographically distributed servers. There are two main approaches of distributed machine learning:

data-parallel approach and model-parallel approach [15, 17]. In the data-parallel approach, the whole data is divided among multiple servers where parallel training takes place for the same machine learning model. In the model-parallel approach, various parameters of a typical machine learning model are trained at distributed servers using exactly the same data. The model-parallel approach can not be used for most of the scenarios where machine learning model parameters can not be split up. Therefore, most of the applications will use a data-parallel approach. Moreover, the data-parallel approach seems more practical because of the different data at distributed locations. Although the data-parallel approach seems more practical for large datasets, it does not primarily address the privacy leakage issue (**C-2**). To address this limitation, federated learning was introduced in [18] that is based on enabling on-device machine learning without transferring the end-devices data to the centralized server for training. Next, we provided a brief overview of federated learning.

1.1.3   Federated Learning Briefing

In federated learning (overview is given in Fig. 1.3), a set of end-devices perform training of local learning models. These local learning models are sent to the centralized aggregation server where aggregation takes place. Then, the aggregation server sends back the global model to end-devices. This process continues in an iterative manner for a number of global federated learning rounds until convergence. It must be noted that federated learning involves iterative interaction between

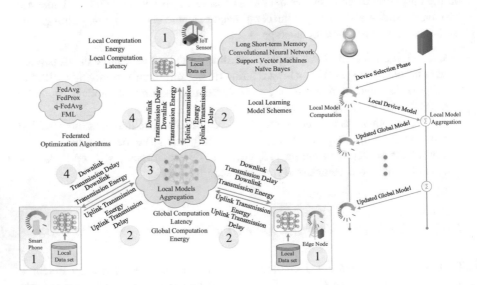

Fig. 1.3 Overview of federated learning

end-devices and the aggregation server. Therefore, to enable efficient interaction, we must employ effective federated optimization scheme. Federated averaging (FedAvg) is based on computing local learning model updates using some machine learning scheme[18]. The choice of the local learning model strictly depends on the target application. We can use convolutional neural network, long-short term memory, support vector machines, and Naive Bayes, among others [19]. Furthermore, to update the local learning model weights, FedAvg uses stochastic gradient descent (SGD). In SGD, the weights are updated at end-devices using partial derivative of loss function w.r.t weights and then multiplied by the learning rate. The learning rate represents the step size of gradient descent. The SGD is given by Khan et al. [15].

$$\omega = \omega - \eta \frac{\partial F}{\partial \omega} \tag{1.1}$$

$$\frac{\partial F}{\partial \omega} \approx \frac{1}{m} \sum_{i \in B} \frac{\partial f_i}{\partial \omega}, \tag{1.2}$$

where m denotes the number of elements in a batch. The computed local learning model weights are averaged by the aggregation server. Finally, the aggregated weights (i.e., global model) is sent back to the end-devices. Although federated learning offer advantages, it faces few challenges. These challenges are privacy issues, computing and communication resource optimization, incentive mechanism, statistical, and system heterogeneity. A malicious aggregation server or end-device can infer other end-devices sensitive information using their learning model updates [16, 20]. Therefore, we must use effective mechanism to truly ensure end-devices privacy preservation. To address privacy leakage issue, one can use differential privacy that adds noise to the end-devices learning model updates before sending it to global aggregation server [21–25]. Other than privacy preservation, there must be some incentive mechanism for attracting the end-devices to participate in federated learning process. Such an incentive mechanism can be either based on game theory, contract theory, or auction theory [26–28]. Furthermore, federated learning involves iterative exchange of learning model updates between the end-devices and aggregation server over a wireless channel. Therefore, one must use efficient wireless resource allocation for federated learning over wireless networks [29–31]. From the aforementioned discussion, we conclude that it is necessary to answer the following questions for federated learning over wireless networks.

- *How does one propose federated optimization (learning) scheme that can efficiently handle statistical and system heterogeneity?*
- *How do we propose an attractive incentive mechanism that will motivate the end-devices to participate in federated learning?*
- *How does one enable efficient resource (i.e., computing and communication resource) optimization to minimize the global convergence time of the global federated learning model?*

- *How do we enable federated learning with enhanced privacy in the presence of a malicious aggregation server?*
- *How do we enable robust federated learning in case of aggregation server malfunctioning due to physical damage or security attack?*

To answer the above questions, we present several proposals in our book. Detailed discussion about the proposals with their chapter details is given in Sect. 1.2.

1.2   Organization of the Book

The goal of this book is to provide a comprehensive study of federated learning for wireless networks. The book consists of three main parts: (a) Fundamentals and background of federated learning for wireless networks, (b) design and analysis of federated learning over wireless networks, and (c) federated learning applications in wireless networks. The first part deals with a brief discussion on the fundamentals of federated learning for wireless networks. In the second part, we comprehensively discuss the design and analysis of wireless federated learning. Specifically, resource optimization, incentive mechanism, security, and privacy are considered. Furthermore, we present several solutions based on optimization theory, graph theory, and game theory to optimize the performance of federated learning over wireless networks. In the final part, we present several applications of federated learning in wireless networks. The book chapters are organized as follows:

- **Chapter 1. Introduction:** In this introductory chapter, we discuss the role of machine learning towards enabling of wireless systems. The key challenges pertaining to implementation of centralized machine learning in wireless systems are also provided. To address the limitations of centralized machine learning, discussions on how to use a distributed machine learning over wireless networks are provided. However, current distributed machine learning schemes do not well tackle the issue of end-devices privacy. Therefore, we present an overview of the fundamental aspects of privacy-preserving federated learning.
- **Chapter 2. Fundamentals of Federated Learning:** This chapter introduces background of machine learning and the key challenges of federated learning. These key challenges are statistical heterogeneity and system heterogeneity. Next, we discuss the key design aspects, such as resource allocation, incentive mechanism, security, and privacy, for federated learning over wireless networks. Finally, we discuss various federated learning algorithms, such as FedAvg, FedProx, q-federated learning, and federated multi-task learning.
- **Chapter 3. Resource Optimization for Wireless Federated Learning:** In this chapter, we present a joint learning and communication framework for federated learning over wireless networks. The formulated optimization problems consider effect of various parameters, such as packet error rate, and transmission latency, on federated learning performance. Specifically, we present wireless resource

allocation and transmit power allocation algorithms for federated learning to optimize its performance. Finally, a novel concept of *dispersed federated learning* is presented. Dispersed federated learning is based on enabling the participation of end-devices with insufficient communication resources in federated learning. Enabling the participation of more end-devices in federated learning will improve the performance of federated learning.

- **Chapter 4. Incentive Mechanisms for Federated Learning:** The purpose of this chapter is to introduce the incentive mechanism design for federated learning. An attractive incentive mechanism is necessary for federated learning to motivate the end-devices participation in federated learning. We present incentive mechanisms based on Stackelberg game and auction theory.
- **Chapter 5. Security and Privacy:** This chapter discusses an efficient secure aggregation method for model updates in federated learning by pre-processing the model updates from each participant and only encrypting portion of the processed updates by functional encryption for inner product to protect the whole parameters, thus achieving efficient aggregation of model update vectors.
- **Chapter 6. Unsupervised Federated Learning:** This chapter considers unsupervised learning tasks being implemented within the federated learning framework to satisfy stringent requirements for low-latency and privacy of the emerging applications. Two DA based unsupervised federated learning schemes are discussed to tackle the problem of non-IID data.
- **Chapter 7. Wireless Virtual Reality:** This chapter introduces the use of federated learning for enabling wireless virtual reality applications. Furthermore, a representative work is presented that focuses on the use of federated learning for the analysis and predictions of orientation and mobility of virtual reality users so as to reduce break in presences of virtual reality users.
- **Chapter 8. Vehicular Networks and Autonomous Driving Cars:** The goal of this chapter is to provide an overview of vehicular networks. Application of federated learning for autonomous driving cars is presented. A dispersed federated learning framework for autonomous cars is proposed. Additionally, an optimization problem is formulated to minimize the dispersed federated learning cost that accounts for transmission latency and packet error rate. To solve the formulated problem, iterative approach is proposed.
- **Chapter 9. Smart Industry and Intelligent Reflecting Surfaces:** This chapter presents various IoT applications of federated learning, such as smart industry and intelligent reflecting surfaces. Representative works for both applications with rigorous problem formulation and solutions are also presented.

Chapter 2
Fundamentals of Federated Learning

Abstract In this chapter, we provide an overview of the fundamentals of FL. First, we discuss a brief history of machine learning. Second, we present the key design challenges for FL over wireless networks. Next, we critically discuss the key design aspects, such as resource allocation, incentive mechanism design, security, and privacy of FL over wireless networks. Finally, we critically discuss existing FL algorithms.

2.1 Introduction and History

The history of machine learning begins in 1943 when the mathematical model of a neural network was first presented by Walter Pitts and Warren McCulloch [32, 33]. Alan Turing proposed a Turing test for testing the intelligence ability of a machine in 1950 [34, 35]. The Turing test is based on an interaction of a machine with humans. A machine will be considered intelligent if it is difficult to distinguish it from humans. Later, the Dartmouth Workshop was organized by John McCarthy, Marvin Minsky, Nathaniel Rochester, and Claude Shannon in 1956 and severed as one of the founding event of artificial intelligence [36]. Lvakhnenko and Lapa in 1965 perhaps presented the first deep neural network [37]. Followed by this, Thomas Cover and Peter E. Hart in 1967 proposed a nearest neighbor pattern algorithm [38]. Fukushima in 1980 proposed a neural network model for a visual pattern recognition [39]. Terrence Sejnowski and Charles Rosenberg proposed an architecture, dubbed as *NETtalk* to construct simplified models for learning human level cognitive tasks in 1986 [40]. Following NETtalk, NETspeak was proposed in 1987 as a reimplementation of NETtalk with further suggestions for performance improvement [41]. In 1986, a generative stochastic artificial neural network, namely, restricted Boltzmann machine (RBM) was presented by Paul Smolensky to learn probability distribution over the set of inputs [32]. Next, many works proposed machine learning algorithms for various applications [32, 42, 43]. Deep learning gained significant interest from the research community since 2006 [44]. Geoffrey E. Hinton in 2009 presented *deep belief networks* which are probabilistic generative models comprising of multiple layers of stochastic, latent variables [45]. On the other hand, recently various projects have been started to propose various popular neural networks. These projects/networks are GoogleBrain (2012), AlexNet (2012), DeepFace(2014), DeepMind (2014), OpenAI (2015), Amazon Machine Learning Platform (2015), ResNet (2015), and U-net (2015) [46].

To train a machine learning model for very large datasets (e.g., astronomical data), we will face the challenges of high computational power consumption and long training time [15]. Additionally, there are many scenarios, where migration of data to a centralized server for training a machine learning model seems difficult [14, 17, 47]. To address these challenges, one can use a distributed machine learning [48–50]. Distributed machine learning trains several models at various geographically distributed servers and combine them to yield a global machine learning model. Distributed learning schemes can be divided into two main types, such as data parallel approach and model parallel approach. The data parallel approach is based on division of data among multiple servers, each server running the same machine learning model. The model parallel approach is based on using the same data for all servers but with a different parameters of a machine learning model. The model parallel approach might not be suitable for various applications because of non-splitting nature of machine learning model. Although distributed machine learning algorithms enable fast learning with less training time by running multiple models on parallel servers, it does not effectively considered the end-devices privacy issues. Migrating the end-devices data to distributed servers will result in end-devices privacy leakage. To address this issue, federated learning was introduced in [18]. Federated learning enables on-device machine learning without moving the end-devices data to a centralized server for training. Although federated learning does not migrate end-devices data to the centralized server, it still has privacy issues [15]. A malicious end-device or aggregation server can infer some of the end-devices private information, and thus federated learning suffers from end-devices privacy leakage. To more effectively ensure privacy in federated learning, one can use various schemes, such as differential privacy and homomorphic encryption. More details about privacy preservation in federated learning are given in Chap. 5. Next, we discuss the federated learning process. Federated learning starts with computing local learning models by the end-devices using their local datasets. After computing the local learning models, all the end-devices send their local models to the aggregation server where aggregation takes place. After global aggregation, the global model is sent back to the end-devices for updating their local models. It must be noted that there exists significant variations in the end-devices system parameters (e.g., CPU-cycles/sec) and local datasets. For an end-device with a certain local dataset and fixed local model iterations, the local model computation time depends on the device's computational capacity (i.e., CPU-cycles/sec). Therefore, different devices will be having different local model computation times for fixed local iterations due to their difference in their local dataset sizes. Additionally, the transmission delay of sending the model parameters will also be different due to wireless channel uncertainties. How much time does an aggregation server wait for receiving local learning model parameters from all the devices? To answer this question, there can be two ways to perform aggregation.

The first one can be a fixed waiting time before a global aggregation takes place. The end-devices with delay greater than the deadline will not be able to participate in the learning process. On the other hand, the aggregation server can use the local models received before the deadline for aggregation while the other local models (i.e., received after the deadline) should be used in the next global aggregation round. Detailed discussions on how to perform federated learning for such scenarios is given below.

Synchronous Federated Learning

In synchronous federated learning, the global model aggregator waits for a fixed amount of time prior to performing aggregation of the end-devices local model updates. All the end-devices involved in synchronous federated learning compute their local models using their local datasets for a fixed time. Next, all the devices send their local model updates to the aggregation server. The aggregation server performs aggregation to yield the global model and send it back all the end-devices. The sequence diagram of the synchronous federated learning is shown in Fig. 2.1. Advantages of the synchronous federated learning are easier management and participation of more devices in learning that will result in improve learning performance [15, 16]. However, this will be at the cost of using significant amount of end-devices computational and communication resources. Enabling all the end-

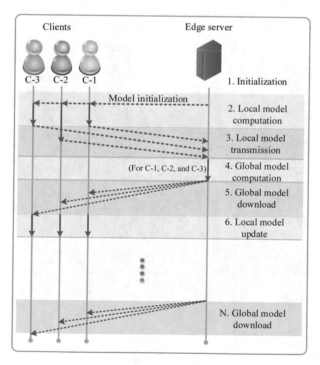

Fig. 2.1 Sequence diagram of synchronous federated learning

devices to finish computation of their local learning models within the deadline, we will face many challenges. There are significant variations in the local dataset sizes. Therefore, end-devices with large local datasets must use more computational resource to compute their local learning models within fixed computation deadline for a particular number of local iterations. Additionally, more communication resources will be required for the end-devices having signal-to-interference-plus-noise-ratio (SINR) values to minimize their transmission delay. The devices whose summation of local model computation delay and transmission latency more than waiting time of the aggregation will not be allowed to participate in the federated learning process. Therefore, how to efficiently use the available communication and end-devices computation resources to enable participation of more end-devices is challenging task in synchronous federated learning.

Asynchronous Federated Learning

In asynchronous federated learning (sequence diagram is shown in Fig. 2.2), there is no fixed deadline for receiving the local learning model updates from end-devices. All the end-devices computes their local learning model updates using local datasets and send the local learning model updates to the aggregation server. The aggregation server performs aggregation of the local learning models from devices. It must be noted here that asynchronous federated learning differs from synchronous

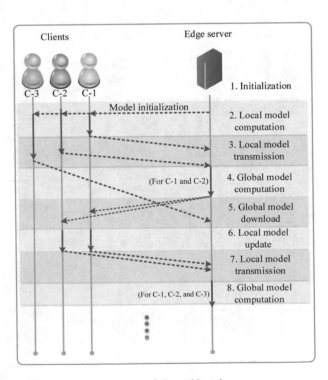

Fig. 2.2 Sequence diagram of asynchronous federated learning

federated learning in deadline for receiving the local learning model updates from end-devices. For practical scenarios, the devices (e.g., smart phone) might be offline some of the time while computing local learning model updates. Generally, increasing the number of local iterations result in performance improvement [18]. Therefore, the devices that remain offline should continue computing their local learning models until they get online. Although devices that remain offline are allowed to participate in federated learning, it will be at the cost of complexity in management of the federated learning process. Furthermore, participation of fewer devices in a global federated learning model computation might result in slow convergence due to the significant data heterogeneity among the end-devices. For instance, consider CIFAR-100 dataset that have 100 classes. If we divide the CIFAR-100 dataset among nodes in such a way that every device should get only 1 class of images. In this scenario, participation of 60 devices out of 100 devices will prolong the convergence time of the global federated learning model. Therefore, it is necessary to make sure the participation of more devices in federated learning process.

2.2 Federated Learning Key Challenges

To implement federated learning over wireless networks, there must be successful interaction between end-devices and aggregation server. At the aggregation server, aggregation (e.g., averaging in case of FedAvg) of local learning model updates takes place. However, aggregation of local learning model updates results in weight divergences. These weight divergences result mainly due to system and statistical heterogeneity. Therefore, one must carefully address the key challenges of statistical and system heterogeneity.

2.2.1 Statistical Heterogeneity

Statistical heterogeneity in federated learning refers to the variations in distributions of the local datasets among the participants of the federated learning. For instance, consider 5 end-devices randomly having the samples of only two classes of the MNIST dataset that has images of 10 different classes. In such a scenario, mostly the devices will be having data from different classes, and thus their data distribution will be different (i.e., statistical heterogeneity). Statistical heterogeneity significantly affect the performance of federated learning. End-devices with statistical heterogeneity will yield different local learning models whose aggregation at the global server might cause weight divergences between the end-devices and aggregation server. FedAvg [18] is based on computing local learning models and simply averaging them to yield the global model without paying significant attention to the data heterogeneity issue. To tackle this issue, FedProx was developed that is based on addition of a weighted proximal term to the local learning model. However, the weight selection for the proximal term in FedProx is challenging. Therefore, there is a need propose efficient federated learning schemes to truly tackle statistical heterogeneity.

2.2.2 System Heterogeneity

System heterogeneity refers to variations in system parameters (e.g., end-devices operating frequencies (CPU-cycles/Sec) and backup energy). These system parameters significantly affect the performance of federated learning. For a synchronous federated learning, all the devices need to compute their local learning model before the computation deadline is reached. However, different end-devices has different local dataset sizes. The devices with less computational resource might not be able to compute their local learning models within the deadline if the number of local iterations are high. Therefore, for devices that have less computational resource, there is a need to run the local model for few iterations. Running a local model for few local iterations will result in generally less global federated learning model accuracy [15, 18]. Therefore, the devices with more computational resources are desirable for participating in federated learning. On the other hand, variations of the wireless channel will significantly degrade the performance of federated learning. Devices having low SINR will suffer from high packet error rates, and thus more performance degradation will occur.

2.3   Key Design Aspects

Overview of federated learning over wireless networks is given in Fig. 2.3. Several machine learning techniques, such as long short-term memory, convolutional neural network, and Naive Bayes schemes can be used at each local device. To enable federated learning, numerous optimization schemes, such as federated averaging (FedAvg) and FedProx can be used to train non-convex federated learning models [51]. FedProx is a modified version of FedAvg that captures both statistical and system heterogeneity among end-devices. FedAvg runs stochastic gradient descent (SGD) on a set of devices to yield local model weights. Subsequently, an averaging of the local weights is performed at the edge computing server located at BS. FedProx has similar steps as FedAvg, but the difference lies in local device minimizing of objective function that considers the objective function of FedAvg with an additional proximal term. By doing so, FedProx limits the impact of non-independent and identically distributed (non-IID) device data on the global learning model. FedAvg does not guarantee theoretical convergence, while FedProx shows theoretical convergence.

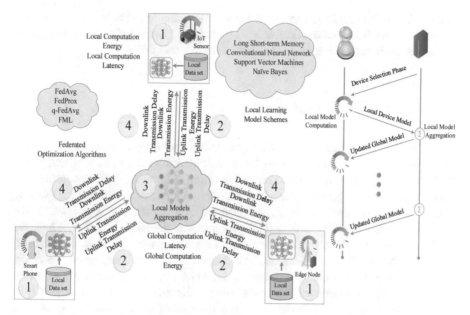

Fig. 2.3 Federated learning sequence diagram

In FedAvg and FedProx, all devices are weighted equally in global federated learning model computation without considering fairness, despite the differences in the device capabilities (e.g., hardware). To capture such fairness among devices, a so-called fairness enabled FedAvg algorithm was proposed [52]. Fairness enabled FedAvg assigns higher weights to devices with poor performance by modifying the objective function of the typical FedAvg algorithm. To introduce potential fairness and reduce training accuracy variance, local devices having a high empirical loss (local loss function) are emphasized by assigning higher relative weight in the fairness enabled FedAvg. Meanwhile, in [53] an adaptive control scheme was proposed to adapt the global federated learning aggregation frequency. This adaptive control scheme offers a desirable tradeoff between global model aggregation and local model update to minimize the loss function with resource budget constraint. All of the above-discussed methods are used for a single task global federated learning model. In real-world IoT systems, it is also of interest to use multi-task federated learning for handling multiple tasks, whose data is distributed among multiple edge nodes. A federated multi-task learning scheme was proposed in [54] by modifying the so-called communication-efficient distributed dual coordinate ascent (CoCoA) framework. To enable a wide variety of machine learning models, CoCoA supports objectives for linear reguarlized loss minimization [55]. In CoCoA, partial results from local computation are effectively combined using optimization problems primal-dual structure. In each round, CoCoA enables the use of any arbitrary optimization algorithm on a local dataset to solve a local learning problem by using distributed optimization for coping with system-level and statistical heterogeneity. To efficiently enable federated optimization schemes over wireless networks, there is a need to address few challenges. These challenges are resource optimization, incentive mechanism design, security, and privacy.

2.3.1   Resource Allocation

Optimization of communication and computation resources is necessary to enable the main phases of federated learning local computation, communication, and global computation. When optimizing federated learning computational and communication resources, the original problem whose goal is to minimize the federated learning cost function can have a dual formulation without constraints. Moreover, if the original problem is convex, then dual problem has the same solution. Thus, the dual problem can be decoupled for obtaining a distributed solution in federated learning. Computation resources can be either those of a local device or of an edge server, whereas communication resources are mainly radio resources of the access network. In the local computation phase, every selected device iteratively performs a local model update using its dataset. The allocation of local device computational resources strongly depends on the device energy consumption, local learning time, and local learning accuracy. Further, the heterogeneity of the local dataset sizes significantly affects the allocation of local computational resources. Device energy consumption and local learning time are strongly dependent on the CPU capability. Increasing the device CPU frequency can increase the energy consumption and decrease the learning time. Similarly, the local computing latency increases for a fixed frequency with an increase in local learning accuracy. Evidently, there is a need to study the tradeoff between computation energy consumption, computational latency, learning time, and learning accuracy. Moreover, the access network and core network resources must be allocated optimally during the communication phase.

2.3.2   Incentive Mechanism

The design of mechanisms that incentivize users to participate in FL is a key challenge. Incentives are possible in different forms, such as user-defined utility and money-based rewards. Several frameworks such as game theory and auction theory can be used in the design of FL incentives [26, 56]. One can design an incentive mechanism using game theory while considering both communication and computation costs. The communication cost can be defined as the total number of rounds used for the interactions between the edge server and end-devices, whereas the computational cost can be the number of local iterations required to compute the local learning model [16]. For synchronous aggregation, given a fixed number of global FL rounds between end-devices and edge server, the convergence rate of the global FL model has a proportional relationship with the number of local iterations. An increase in the number of local iterations minimizes the local learning model error and thus, few global FL rounds are required to reach a certain global FL model accuracy. Therefore, for a fixed global FL model accuracy, an increase in computational cost reduces communication cost and vice versa. For instance, consider a incentive mechanism game whose players are the edge server and edge users. The edge server announces a reward as an incentive to the participating users while maximizing its benefits in terms of improving global FL model accuracy. Meanwhile, the edge users maximize their individual utilities to improve their benefit. One example of a user utility could be the improvement of local learning model accuracy within the allowed communication time during FL training. An improvement in the local learning model accuracy of the end-user increases its incentive from the edge server and vice versa. This process of incentive-based sharing of model parameters continues until convergence to some global model accuracy level.

2.3.3 Security and Privacy

Security in federated learning over wireless networks can be divided into devices physical security and cyber security. Devices physical security refers to physical access of the end-devices, distributed edge servers and cloud server used for aggregation of local models. Cyber security deals with end-devices, edge servers, and the cloud server authentication. Moreover, it takes into account the wireless security during exchange of learning model parameters between the end-devices and aggregation server. It is very difficult to restrict the physical access of the end-devices and edge servers due to their geographically distributed nature. Therefore, one must propose novel and light weight authentication schemes for end-devices, edge servers, and the cloud servers. Generally, the computational complexity of authentication scheme increases by raising its protection against the security attacks. Therefore, depending on the available computing power, one must design authentication scheme. There is a need to propose light-weight authentication with low computational complexity for end-devices, whereas edge aggregation servers can be installed with authentication schemes of higher complexity because of their higher computing power than end-devices. Meanwhile, for a cloud server, one can propose authentication scheme with the highest complexity due to more available computing power at the cloud among the three players (i.e., end-devices, edge server, and cloud server).

Federated learning involves iterative exchange of learning model updates between the end-devices and aggregation servers using wireless channel. Therefore, an attacker can easily access the learning model parameters. Accessing the local learning model parameters will results two kind of issues. First one deals with the end-devices sensitive information leakage, whereas the second one deals with altering the learning model parameters before aggregation. This alteration of the local learning model will prolong the global federated learning model convergence.

Another possible impact can either the global federated learning model does not converge if the attacker continuously perform malicious activity. Therefore, one must use effective encryption schemes to avoid attacker from accessing and altering the local learning models.

2.4 Federated Learning Algorithms

Consider a set $\mathcal{N}$ of N devices, each is with local dataset $\mathcal{D}_n$ of D_n data points. The goal of the federated learning is to minimize the overall loss function.

$$\underset{w_1, w_2, \ldots, w_N}{\text{minimize}} \frac{1}{K} \sum_{n=1}^{N} \sum_{k=1}^{k_n} f_{\text{FL}}(\boldsymbol{w}_n, d_{nk}, \Theta_{nk}), \qquad (2.1a)$$

$$s.t. \boldsymbol{w}_1 = \boldsymbol{w}_2 = \ldots = \boldsymbol{w}_N = z, \forall n \in \mathcal{N}, \qquad (2.1b)$$

where K denotes the total data points of all the N devices involved in federated learning process. f_{FL} is the loss function that is problem dependent. For instance, f_{FL} for regression problem can be given by $f_{\text{FL}}(\boldsymbol{w}_n, d_{nk}, \Theta_{nk}) = \frac{1}{2}(d_{nk} w_n - \Theta_{nk})^2$.

2.4.1 FedAvg

FedAvg was introduced in [18] to enable distributed, on-device machine learning without transferring the end-devices data to the centralized server. The main focus of the work in [18] was to tackle the devices system and statistical heterogeneity. FedAvg is based on modification of stochastic gradient descent (SGD). SGD involves a single batch gradient calculation on a randomly selected client per communication round, and thus suffers from large number of communication rounds between the end-devices and aggregation server. We can extend SGD for federated settings. Federated SGD involves computation of local learning models using single iteration, that are aggregated to yield the global model. To further improve the performance of the federated SGD, there is a need to perform multiple local iterations at all the devices before a global aggregation takes place. For a fixed global federated learning accuracy, performing multiple local aggregations at end-devices will reduce the number of communication rounds and vice versa. The summary of the FedAvg algorithm is given in Algorithm 1. A set of end-devices limited by a fraction C among all devices is selected to train their local models. These end-devices compute their local learning models for a fixed number of local iterations and send their local models to the aggregation server. The aggregation server takes average of all the received local model updates and send back the global model to all devices for updating their local models.

Algorithm 1 FedAvg [18]

 1: ***Aggregation Server***
 2: Weights initialization ω^0
 3: **for** t=0, 1,..., Global Rounds-1 **do**
 4:     Select $\mathcal{N}_s \leftarrow m = max(C.N)$ clients randomly.
 5:     **for** For every device $n \in \mathcal{N}_s$, parallel run. **do**
 6:         $\omega_n^{t+1} \leftarrow DeviceUpdate(n, \omega_n^t)$
 7:         $\omega^{t+1} \leftarrow \sum_{n=1}^{N_s} \frac{k_n}{K} \omega_n^{t+1}$
 8:     **end for**
 9: **end for**
10: ***DeviceUpdate(k,ω)***
11: $\mathcal{B} \leftarrow$ Split d_k into batches.
12: **for** e=0, 1, .., Local iterations-1 **do**
13:     **for** $b \in \mathcal{B}$ **do**
14:         $\omega \leftarrow \omega - \eta \triangledown l(\omega, b)$
15:     **end for**
16: **end for**

2.4.2 FedProx

Although FedAvg has shown empirical success in heterogeneous settings, it still has few limitations [15]. FedAvg does not fully address the heterogeneity. For instance, end-devices are not allowed to perform variable amount of training work (i.e., local iterations). Instead, end-devices in FedAvg with less computational resource that fails to meet the local model computation deadline are not allowed to participate in federated learning process. According to [18], an increase in number of local iterations will reduce global rounds for attaining a certain global accuracy. However, performing more local iterations for heterogeneous devices may cause the global federated learning model to diverge. Therefore, one must address this issue of devices heterogeneity for federated learning. Furthermore, FedAvg does not provably guarantee theoretical convergence for many practical scenarios [51]. Therefore, it is necessary to propose new federated learning scheme. FedProx was presented in [57] to account more effectively for the heterogeneity of the federated learning system. FedProx is based on addition of a proximal term in local learning model to account more better for the heterogeneity among nodes.

$$\min_{\omega} h_n(\omega; \omega^t) = F_n(\omega) + \frac{\mu}{2} \parallel \omega - \omega^t \parallel^2 \qquad (2.2)$$

The addition of a proximal term has two advantages: (a) addresses the devices heterogeneity issue by restricting the local model to closer to the global model without the need of manual adjustment using local iteration settings, and (b) it offers incorporation of variable amount of training works because of the system heterogeneity. The summary of the FedProx algorithm is shown in Algorithm 2.

Algorithm 2 FedProx [57]

1: *Aggregation Server*
2: Weights initialization ω^0
3: **for** t=0, 1,..., Global Rounds-1 **do**
4:      Select $\mathcal{N}_s \leftarrow m = max(C.N)$ clients randomly.
5:      **for** For every device $n \in \mathcal{N}_s$, parallel run. **do**
6:          Compute ω_n^{t+1} which is a γ_n^t-inexact minimizer of: $\omega_n^{t+1} \approx \min_{\omega} h(\omega; \omega^t) = F_n(\omega) + \frac{\mu}{2} \| \omega - \omega^t \|^2$
7:          $\omega^{t+1} \leftarrow \sum_{n=1}^{N_s} \frac{k_n}{K} \omega_n^{t+1}$
8:      **end for**
9: **end for**

2.4.3 q-Federated Learning

FedAvg [18] was based on simple averaging of the local learning model updates at the aggregation server that might not converge fast for heterogeneous conditions (i.e., system and statistical heterogeneity). To address these issues of FedAvg, FedProx was introduced in [57] that is based on addition of a proximal term to the local learning model to minimize deviation of the local learning from the global model during the local update step. However, both FedAvg and FedProx do not take into account the resource fairness issues. For instance, some of the end-device might not get enough communication resources, and thus will suffer from performance degradation. Many resource allocation algorithms in literature focused on maximizing the overall throughput that might result in allocation of less resources to few nodes. On the other hand, fairness in federated learning can be the uniformity of accuracy distribution among end-devices. q-fair federated learning was introduced in [52] to account for the poor performing end-devices. In contrast to FedAvg [18], q-fair federated learning uses weighted averaging. To introduce potential fairness and reduce training accuracy variance, local devices having a high empirical loss (local loss function) are emphasized by assigning higher relative weight in the q-federated learning.

2.4.4 Federated Multi-Task Learning

Federated learning multi-task learning was proposed in [54] to enable training of multiple models for various nodes. More specifically, the reason for training the multiple models for different nodes was to tackle the statistical heterogeneity. Federated multi-task learning was based on modifying the so-called communication-efficient distributed dual coordinate ascent (CoCoA) framework. To enable a wide variety of machine learning models, CoCoA supports objectives for linear reguarlized loss minimization [55]. In CoCoA, partial results from local computation are effectively combined using optimization problems primal-dual structure. In each round, CoCoA enables the use of any arbitrary optimization algorithm on a local dataset to solve a local learning problem by using distributed optimization for coping with system-level and statistical heterogeneity.

2.5   Summary

In this chapter, we have discussed the evolution of artificial intelligence. The key challenges of federated learning, such as statistical heterogeneity and system heterogeneity, are presented. Next, the key design aspects of federated learning over wireless networks, such as resource allocation, incentive mechanism design, and security and privacy, are presented. Finally, we presented an overview of various federated learning algorithms, such as FedAvg, FedProx, q-federated learning, and federated multi-task learning.

Part II
Wireless Federated Learning: Design and Analysis

Chapter 3
Resource Optimization for Wireless Federated Learning

Abstract This chapter provides an overview of various resources such as computational, communication, and power resources, required for wireless federated learning. We perform convergence analysis of wireless federated learning. Additionally, joint resource and power allocation for wireless federated learning are proposed. Finally, we present a collaborative federated learning framework to efficiently enable the participation of communication-resource deficient devices in the federated learning process for performance improvement.

3.1 Introduction

There are two main aspects of federated learning over wireless networks, such as wireless for federated learning and federated learning for wireless networks. Wireless for federated learning deals with the optimization of wireless communication resources for implementing federated learning, whereas federated learning for wireless deals with using federated learning for implementing various intelligent network functions (e.g., intelligent edge caching, intelligent transceivers, intrusion detection, vehicular communication). In this chapter, we focus on the wireless for federated learning aspect. Federated learning over wireless networks mainly involves four steps: (a) local model computation, (b) transfer of local learning model updates from end-devices to the aggregation server, (c) aggregation of local models, and (d) downloading the global model updates, as shown in Fig. 3.1. To carry out the above four steps, federated learning uses two kinds of resources, such as computing resources and wireless resources. The computing resource can be either local (i.e., at the end-devices) or global (i.e., aggregation server). At the end-devices, the computing resource such as CPU-cycles/sec determines the local learning model computation time for a fixed local dataset size and local learning model architecture. However, devices' energy consumption increases with an increase in the local computing resource. Therefore, it is necessary to make a tradeoff between the local model computing time and energy consumption. On the other hand, federated learning involves iterative interaction between end-devices and the aggregation over a wireless channel. Training of a massive number of end-devices using

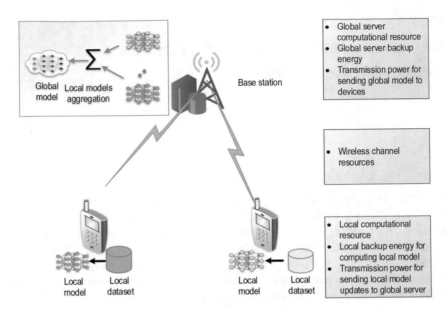

Fig. 3.1 An overview of resources required for wireless federated learning

federated learning will consume a significant amount of communication resources. Therefore, one must efficiently perform resource allocation for federated learning over wireless networks. Furthermore, there might be a possibility of end-devices not participating in the federated learning process due to insufficient communication resources. To address this limitation, one can use collaborative federated learning. In collaborative federated learning, end-devices with insufficient communication resources sending their local learning model updates to their nearby devices where local aggregation takes place. Then, the locally aggregated models are sent to the global aggregation server. Collaborative federated learning will result in significant performance improvement. Collaborative federated learning will be explained in more detail later in this chapter.

3.2   Wireless Federated Learning: Convergence Analysis and Resource Allocation

Although federated Learning enables on-device machine learning, it has few challenges. In contrast to centralized machine learning, global federated learning computation time includes not only the end-device computation times but also the communication time for transferring learning model updates between end-devices and the aggregation server. The computational time depends on end-devices' computational capacity (i.e., CPU-cycles/sec) and local data sizes. Moreover, communication times primarily depend on end-devices channel gains and learning model update size. To deploy federated learning over wireless networks, we must answer the following two questions:

- *When UEs should spend more time on local model computation to achieve high learning accuracy and less communication updates, or vice versa?*
- *How to strike a balance between two conflicting goals of minimizing federated Learning time and end-device energy consumption?*

To address these questions, we present "Federated Learning over Wireless Networks" problem design and analysis, which can be summarized as follows:

- We present federated learning over wireless networks problem that captures two trade-offs: (i) learning time versus end-device energy consumption by using the Pareto efficiency model, and (ii) computation versus communication learning time by finding the optimal learning accuracy parameter.
- Despite the non-convex nature of the formulated problem, we exploit its special structure and use the variable decomposition approach to split and transform formulated problem into three convex sub-problems. We show that the first two sub-problems can be solved separately, then their solutions are used to obtain the solution to the third sub-problem. By analyzing the closed-form solution to each sub-problem, we obtain qualitative insights into the impact of the Pareto-efficient controlling knob to the optimal: (i) computation and communication learning time, (ii) UE resource allocation, and (iii) learning accuracy. Finally, the combined solution to all sub-problems can provide the globally optimal solution to main formulated problem.
- We further provide extensive numerical results to examine the: (i) impact of UE heterogeneity, (ii) Pareto curve between UE energy cost and system learning time, and (iii) the impact of the proportion of computation over communication time on the optimal accuracy level.

3.2.1   System Model

We consider a wireless multi-user system consisting of one base station (BS) and a set $\mathcal{N}$ of N UEs. Each participating UE n stores a local data set $\mathcal{D}_n$, with its size that is denoted by D_n. Then, we can define the total data size by $D = \sum_{n=1}^{N} D_n$. In an example of the supervised learning setting, at UE n, $\mathcal{D}_n$ defines the collection of data samples given as a set of input-output pairs $\{x_i, y_i\}_{i=1}^{D_n}$, where $x_i \in \mathbb{R}^d$ is an input sample vector with d features, and $y_i \in \mathbb{R}$ is the labeled output value for the sample x_i. The data can be generated through the usage of UE, for example, via interactions with mobile apps. With these UEs data, several machine learning applications can be employed for wireless networks such as predicting the BS's load in the next hours for dynamic BS load balancing, or predicting the next hovering position of drones so that their coverage is optimized.

In a typical learning problem, for a sample data $\{x_i, y_i\}$ with input x_i (e.g., the response time of various apps inside the UE), the task is to find the *model parameter* w that characterizes the output y_i (e.g., label of BS load, such as high or low, in next hours) with the loss function $f_i(w)$. Some examples of the loss function are $f_i(w) = \frac{1}{2}(x_i^T w - y_i)^2$, $y_i \in \mathbb{R}$ for linear regression and $f_i(w) = \{0, 1 - y_i x_i^T w\}$, $y_i \in \{-1, 1\}$ for support vector machine. Hence, the loss function on the data set of UE n is defined as

$$J_n(w) := \frac{1}{D_n} \sum_{i \in \mathcal{D}_n} f_i(w). \tag{3.1}$$

Then, the learning model is the minimizer of the following global loss function minimization problem

$$\min_{w \in \mathbb{R}^d} J(w) := \sum_{n=1}^{N} \frac{D_n}{D} J_n(w). \tag{3.2}$$

Federated Learning Over Wireless Networks

In this section, we adapt the Federated Learning framework [58] to the wireless networks, namely, FEDL, as the following.

1. At the UE side, there are two phases at tth update:
   Computation. Each UE n solves its local problem

   $$w_n^{(t)} = \arg\min_{w_n \in \mathbb{R}^d} F_n\left(w_n \mid w^{(t-1)}, \nabla J^{(t-1)}\right) \tag{3.3}$$

   with a local accuracy[1] $0 \leq \theta \leq 1$ (i.e., $||\nabla F_n(w^{(t)})|| \leq \theta \, ||\nabla F_n(w^{(t-1)})||$).
   Communication. All UEs share the wireless environment to transmit $w_n^{(t)}$ and the gradient $\nabla J_n^{(t)}$ to BS.
2. At the BS, the following information is aggregated

   $$w^{(t+1)} = \frac{1}{N} \sum_{n=1}^{N} w_n^{(t)} \tag{3.4}$$

   $$\nabla J^{(t+1)} = \frac{1}{N} \sum_{n=1}^{N} \nabla J_n^{(t)} \tag{3.5}$$

   and fed-back to all UEs. This process is iterative until a global accuracy $0 \leq \varepsilon \leq 1$ is achieved (i.e., $||\nabla J(w^{(t)})|| \leq \varepsilon ||\nabla J(w^{(t-1)})||$).

---

[1] Here $\theta = 0$ means the local problem is required to be solved optimally, and $\theta = 1$ means no progress for local problem [58].

To solve problem (3.2), **FEDL** uses an iterative approach that requires a number of *global iterations* (i.e., communication rounds) to achieve a global accuracy level ε. In each global iteration, there are interactions between the UEs and BS. Specifically, a participating UE, in each computation phase, will minimize its objective $F_n(w_n)$ in (3.3) using local training data $\mathcal{D}_n$. Minimizing F_n also takes multiple *local iterations* up to an accuracy threshold θ that is common to all UEs. As in [59], the computation phase is *synchronous* such that all UEs have to finish solving their local problems before entering the communication phase to transmit their updates to the BS by using a wireless medium sharing scheme (e.g., time-sharing similar to TDMA).

The BS then aggregates the local model parameters and gradients, i.e., w_n and the ∇J_n, $\forall n$, respectively, to update and then broadcast the global model parameters and gradients, i.e., w and the ∇J according to (3.4) and (3.5), respectively, which are required for participating UEs to minimize their $F_n(w_n)$,[2] in the next global iteration. We see that the BS does not access the local data $\mathcal{D}_n$, $\forall n$, hence *preserving data privacy*.

For strongly convex objective $J(w)$, the global iterations is shown to be [59]

$$K(\varepsilon, \theta) = \frac{\mathcal{O}(\log(1/\epsilon))}{1 - \theta}, \qquad (3.6)$$

which is affected by both global accuracy ε and local accuracy θ. For example, when ε and θ are small (more accurate), **FEDL** needs to runs more global iterations. On the other hand, each global iteration consists of both computation and uplink communication time. Since the downlink bandwidth is larger than that of uplink and the BS power is much higher than UE's transmission power, the downlink time is negligible compared to the uplink time and thus is not considered in this work. The computation time, however, depends on the number of local iterations, which is upper bounded by $\mathcal{O}(\log(1/\theta))$ for a wide range of iterative algorithms to solve (3.3) such as gradient descent, coordinate descent, or stochastic dual coordinate descent [60]. In [59], it is shown that **FEDL** performance does not depend on which algorithms are used in the computation phase as long as the convergence time of that algorithm is upper-bounded by $\mathcal{O}(\log(1/\theta))$. Denote the time of one local iteration by T_{cmp}, then the computation time in one global iteration is $v \log(1/\theta)T_{cmp}$ for some positive constant v that depends on the data size and condition number of the local problem [60]. Denoting the communication time in one global iteration by T_{com}, the total time of one global iteration of **FEDL** is defined as

$$T_{glob}(T_{cmp}, T_{com}, \theta) := T_{com} + v \log(1/\theta)T_{cmp}. \qquad (3.7)$$

---

[2] One example, from [58], is $F_n(w_n) = J_n(w_n) - \left(\nabla J_n^{(t-1)} - \beta_1 \nabla J^{(t-1)}\right)^T w_n + \frac{\beta_2}{2}||w_n - w_n^{(t-1)}||^2$ where $\beta_1, \beta_2 \geq 0$ are parameters.

In this work, we consider a fixed global accuracy ϵ, so we normalize $\mathcal{O}(\log(1/\epsilon))$ to 1 so that $K(\theta) = \frac{1}{1-\theta}$ for ease of presentation. Furthermore, we also normalize v to 1 since we can absorb v into T_{cmp} as the upper bound of one local computation iteration. Thus the upper-bound of FEDL learning time is $K(\theta) T_{glob}(\theta)$. Henceforth, we omit the word "upper-bound" for brevity. In the next sub-sections, we will present how computation and communication time relate to UEs' energy consumption.

Computation Model

We denote the number of CPU cycles for UE n to execute one sample of data by c_n, which can be measured offline [61] and is known as a priori. Since all samples $\{x_i, y_i\}_{i \in \mathcal{D}_n}$ have the same size (i.e., number of bits), the number of CPU cycles required for UE n to run one local iteration is $c_n D_n$. Denote the CPU-cycle frequency of the UE n by f_n. Then the CPU energy consumption of UE n for one local iteration of computation can be expressed as follows [62]

$$E_n^{cmp}(f_n) = \sum_{i=1}^{c_n D_n} \frac{\alpha_n}{2} f_n^2 = \frac{\alpha_n}{2} c_n D_n f_n^2, \tag{3.8}$$

where $\alpha_n/2$ is the effective capacitance coefficient of UE n's computing chipset. Furthermore, the computation time per local iteration of the UE n is $\frac{c_n D_n}{f_n}$, $\forall n$. We denote the vector of f_n by $f \in \mathbb{R}^n$.

Communication Model

In FEDL, regarding the communication phase of UEs, we consider a time-sharing multi-access protocol for UEs. We note that this time-sharing model is not restrictive because other schemes, such as OFDMA, can also be applied to FEDL. The achievable transmission rate (nats/s) of UE n is defined as follows:

$$r_n = B \ln\left(1 + \frac{h_n p_n}{N_0}\right), \tag{3.9}$$

where B is the bandwidth, N_0 is the background noise, p_n is the transmission power, and h_n is the channel gain of the UE n. We assume that h_n is constant during the learning time of FEDL.[3] Denote the fraction of communication time allocated to UE n by τ_n, and the data size (in nats) of both w_n and ∇J_n by s_n. Because the

---

[3] We treat the case of random h_n by adding the outage probability constraint, e.g., for Rayleigh fading channel, $\Pr\left(\frac{h_n p_n}{N_0} < \gamma\right) \leq \zeta$ where γ is the SNR threshold and ζ is the bounded probability [63]. This constraint is equivalent to $p_n \geq \frac{-\gamma N_0}{\log(1-\zeta)}$ and can be integrated to the constraint (3.17) without changing any insights of the considered problem.

dimensions of vectors w_n and ∇J_n are fixed, we assume that their sizes are constant throughout the FEDL learning. Then the transmission rate of each UE n is

$$r_n = s_n/\tau_n, \tag{3.10}$$

which is shown to be the most energy-efficient transmission policy [64]. Thus, to transmit s_n within a time duration τ_n, the UE n's energy consumption is

$$E_n^{com}(\tau_n) = \tau_n p_n = \tau_n \, p_n(s_n/\tau_n), \tag{3.11}$$

where the power function is

$$p_n(s_n/\tau_n) := \frac{N_0}{h_n}\left(e^{\frac{s_n/\tau_n}{B}} - 1\right) \tag{3.12}$$

according to (3.9) and (3.10). We denote the vector of τ_n by $\tau \in \mathbb{R}^n$.

3.2.2 Problem Formulation

Define the total energy consumption of all UEs for each global iteration by E_{glob}, which is expressed as follows:

$$E_{glob}(f, \tau, \theta) := \sum_{n=1}^{N} E_n^{com}(\tau_n) + \log(1/\theta) E_n^{cmp}(f_n).$$

Then, we consider an optimization problem, abusing the same name FEDL, as follows

$$\text{FEDL:} \quad \min. \quad K(\theta)\big[E_{glob}(f, \tau, \theta) + \kappa \, T_{glob}(T_{cmp}, T_{com}, \theta)\big] \tag{3.13}$$

$$\text{s.t.} \quad \sum_{n=1}^{N} \tau_n \leq T_{com}, \tag{3.14}$$

$$\max_n \frac{c_n D_n}{f_n} = T_{cmp}, \tag{3.15}$$

$$f_n^{min} \leq f_n \leq f_n^{max}, \ \forall n \in \mathcal{N}, \tag{3.16}$$

$$p_n^{min} \leq p_n(s_n/\tau_n) \leq p_n^{max}, \ \forall n \in \mathcal{N}, \tag{3.17}$$

$$0 \leq \theta \leq 1. \tag{3.18}$$

To minimize both UEs' energy consumption and the Federated Learning time are conflicting. For example, the UEs can save energy by setting the lowest frequency level all the time, but this will certainly increase the learning time. Therefore, to strike the balance between energy cost and learning time, the weight

κ (Joules/second), used in the objective as an amount of additional energy cost that **FEDL** is willing to bear for one unit of learning time to be reduced, captures the Pareto-optimal tradeoff between the UEs' energy cost and the Federated Learning time. For example, when most of the UEs are plugged in, then UE energy cost is not the main concern, thus κ can be large.

While constraint (3.14) captures the time-sharing uplink transmission of UEs, constraint (3.15) defines that the computing time in one local iteration is determined by the "bottleneck" UE (e.g., with large data size and low CPU frequency). The feasible regions of CPU-frequency and transmit power of UEs are imposed by constraints (3.16) and (3.17), respectively. We note that (3.16) and (3.17) also capture the heterogeneity of UEs with different types of CPU and transmit chipsets. The last constraint restricts the feasible range of the local accuracy.

3.2.3   Decomposition-Based Solution

We see that **FEDL** is non-convex due to the constraint (3.15) and several products of two functions in the objective function. However, in this subsection, we will characterize its optimal solution by decomposing it into multiple convex sub-problems.

We consider the first case when θ is fixed, then **FEDL** can be decomposed into two sub-problems as follows:

$$\textbf{SUB1:} \quad \min. \qquad \sum_{n=1}^{N} E_n^{cmp}(f_n) + \kappa T_{cmp} \tag{3.19}$$

$$\text{s.t.} \qquad \frac{c_n D_n}{f_n} \leq T_{cmp}, \ \forall n \in \mathcal{N}, \tag{3.20}$$

$$f_n^{min} \leq f_n \leq f_n^{max}, \ \forall n \in \mathcal{N}. \tag{3.21}$$

and

$$\textbf{SUB2:} \quad \min. \qquad \sum_{n=1}^{N} E_n^{com}(\tau_n) + \kappa T_{com} \tag{3.22}$$

$$\text{s.t.} \qquad \sum_{n=1}^{N} \tau_n \leq T_{com}, \tag{3.23}$$

$$p_n^{min} \leq p_n(s_n/\tau_n) \leq p_n^{max}, \ \forall n \in \mathcal{N}. \tag{3.24}$$

While **SUB1** is a CPU-cycle control problem for the computation time and energy minimization, **SUB2** can be considered as an uplink power control to determine the UEs' fraction of time sharing to minimize the UEs energy and communication time. We note that the constraint (3.15) of **FEDL** is replaced by an equivalent one (3.20) in **SUB1**. We can consider T_{cmp} and T_{com} as virtual deadlines for UEs to perform their computation and communication updates, respectively.

It can be observed that both **SUB1** and **SUB2** are convex problems. We note that the constant factors $K(\theta)\log(1/\theta)$ and $K(\theta)$ of **SUB1** and **SUB2**'s objectives, respectively, are omitted since they have no effects on these sub-problems' solutions.

SUB1 Solution

We first propose Algorithm 3 in order to categorize UEs into one of three groups: $\mathcal{N}_1$ is a group of "bottleneck" UEs that always run its maximum frequency, $\mathcal{N}_2$ is the group of "strong" UEs which can finish their tasks before the computational virtual deadline even with the minimum frequency, and $\mathcal{N}_3$ is the group of UEs having the optimal frequency inside the interior of their feasible sets.

Lemma 3.1 *The optimal solution to* **SUB1** *is as follows*

$$f_n^* = \begin{cases} f_n^{max}, & \forall n \in \mathcal{N}_1, \\ f_n^{min}, & \forall n \in \mathcal{N}_2, \\ \frac{c_n D_n}{T_{cmp}^*}, & \forall n \in \mathcal{N}_3, \end{cases} \tag{3.25}$$

$$T_{cmp}^* = \max\{T_{\mathcal{N}_1}, T_{\mathcal{N}_2}, T_{\mathcal{N}_3}\}, \tag{3.26}$$

where $\mathcal{N}_1, \mathcal{N}_2, \mathcal{N}_3 \subseteq \mathcal{N}$ *are three subsets of UEs produced by Algorithm 3 and*

$$T_{\mathcal{N}_1} = \max_{n \in \mathcal{N}} \frac{c_n D_n}{f_n^{max}} \tag{3.27}$$

$$T_{\mathcal{N}_2} = \max_{n \in \mathcal{N}_2} \frac{c_n D_n}{f_n^{min}} \tag{3.28}$$

$$T_{\mathcal{N}_3} = \left(\frac{\sum_{n \in \mathcal{N}_3} \alpha_n (c_n D_n)^3}{\kappa} \right)^{1/3}. \tag{3.29}$$

From Lemma 3.1, first, we see that the optimal solution depends not only on the existence of these subsets, but also on their virtual deadlines $T_{\mathcal{N}_1}$, $T_{\mathcal{N}_2}$, and $T_{\mathcal{N}_3}$, in which the longest of them will determine the optimal virtual deadline T_{cmp}^*. Second, from (3.25), the optimal frequency of each UE will depend on both T_{cmp}^* and the subset it belongs to. We note that depending on κ, some of the three sets (not all) are possibly empty sets, and by default $T_{\mathcal{N}_i} = 0$ if $\mathcal{N}_i$ is an empty set, $i = 1, 2, 3$. Next, by varying κ, we observe the following special cases.

Algorithm 3 Finding $\mathcal{N}_1, \mathcal{N}_2, \mathcal{N}_3$ in Lemma 3.1

1: Sort UEs such that $\frac{c_1 D_1}{f_1^{min}} \leq \frac{c_2 D_2}{f_2^{min}} \cdots \leq \frac{c_N D_N}{f_N^{min}}$
2: **Input:** $\mathcal{N}_1 = \emptyset, \mathcal{N}_2 = \emptyset, \mathcal{N}_3 = \mathcal{N}, T_{\mathcal{N}_3}$ in (3.29)
3: **for** $i = 1$ to N **do**
4:    **if** $\max_{n \in \mathcal{N}} \frac{c_n D_n}{f_n^{max}} \geq T_{\mathcal{N}_3} > 0$ and $\mathcal{N}_1 == \emptyset$ **then**
5:        $\mathcal{N}_1 = \mathcal{N}_1 \cup \{m : \frac{c_m D_m}{f_m^{max}} = \max_{n \in \mathcal{N}} \frac{c_n D_n}{f_n^{max}}\}$
6:        $\mathcal{N}_3 = \mathcal{N}_3 \setminus \mathcal{N}_1$ and update $T_{\mathcal{N}_3}$ in (3.29)
7:    **end if**
8:    **if** $\frac{c_i D_i}{f_i^{min}} \leq T_{\mathcal{N}_3}$ **then**
9:        $\mathcal{N}_2 = \mathcal{N}_2 \cup \{i\}$
10:       $\mathcal{N}_3 = \mathcal{N}_3 \setminus \{i\}$ and update $T_{\mathcal{N}_3}$ in (3.29)
11:   **end if**
12: **end for**

Corollary 3.1 *The optimal solution to* **SUB1** *can be divided into four regions as follows.*

(a) $\kappa \leq \min_{n \in \mathcal{N}} \alpha_n (f_n^{min})^3$:
    $\mathcal{N}_1$ and $\mathcal{N}_3$ are empty sets. Thus, $\mathcal{N}_2 = \mathcal{N}$, $T_{com}^* = T_{\mathcal{N}_2} = \max_{n \in \mathcal{N}} \frac{c_n D_n}{f_n^{min}}$, and
    $f_n^* = f_n^{min}, \forall n \in \mathcal{N}$.

(b) $\min_{n \in \mathcal{N}} \alpha_n (f_n^{min})^3 < \kappa \leq \left(\max_{n \in \mathcal{N}_2} \frac{c_n D_n}{f_n^{min}}\right)^3$:
    $\mathcal{N}_2$ and $\mathcal{N}_3$ are non-empty sets, whereas $\mathcal{N}_1$ is empty. Thus, $T_{cmp}^* = \max\{T_{\mathcal{N}_2}, T_{\mathcal{N}_3}\}$, and $f_n^* = \max\{\frac{c_n D_n}{T_{cmp}^*}, f_n^{min}\}, \forall n \in \mathcal{N}$.

(c) $\left(\max_{n \in \mathcal{N}_2} \frac{c_n D_n}{f_n^{min}}\right)^3 < \kappa \leq \frac{\sum_{n \in \mathcal{N}_3} \alpha_n (c_n D_n)^3}{\left(\max_{n \in \mathcal{N}} \frac{c_n D_n}{f_n^{max}}\right)^3}$:
    $\mathcal{N}_1$ and $\mathcal{N}_2$ are empty sets. Thus $\mathcal{N}_3 = \mathcal{N}$, $T_{cmp}^* = T_{\mathcal{N}_3}$, and $f_n^* = \frac{c_n D_n}{T_{\mathcal{N}_3}}, \forall n \in \mathcal{N}$.

(d) $\kappa > \frac{\sum_{n \in \mathcal{N}_3} \alpha_n (c_n D_n)^3}{\left(\max_{n \in \mathcal{N}} \frac{c_n D_n}{f_n^{max}}\right)^3}$:
    $\mathcal{N}_1$ is non-empty. Thus $T_{cmp}^* = T_{\mathcal{N}_1}$, and

$$
f_n^* = \begin{cases} f_n^{max}, & \forall n \in \mathcal{N}_1 \\ \max\{\frac{c_n D_n}{T_{\mathcal{N}_1}}, f_n^{min}\}, & \forall n \in \mathcal{N} \setminus \mathcal{N}_1 \end{cases} \tag{3.30}
$$

We illustrate Corollary 3.1 in Fig. 3.2 with four regions[4] as follows.

---

[4] All closed-form solutions are also verified by the solver IPOPT [65].

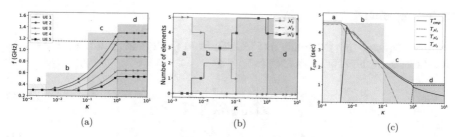

Fig. 3.2 Solution to **SUB1** with five UEs. For wireless communication model, the UE channel gains follow the exponential distribution with the mean $g_0(d_0/d)^4$ where $g_0 = -40\,$dB and the reference distance $d_0 = 1\,$m. The distance between these devices and the wireless access point is uniformly distributed between 2 and 50 m. In addition, $B = 1$ MHz, $\sigma = 10^{-10}$ W, the transmission power of devices are limited from 0.2 to 1 W. For UE computation model, we set the training size D_n of each UE as uniform distribution in 5–10 MB, c_n is uniformly distributed in 10–30 cycles/bit, f_n^{max} is uniformly distributed in 1.0–2.0 GHz, $f_n^{min} = 0.3$ GHz. Furthermore, $\alpha = 2 \times 10^{-28}$ and the UE update size $s_n = 25{,}000$ nats (≈ 4.5 KB). (**a**) Optimal CPU frequency of each UE. (**b**) Three subsets outputted by Algorithm 3. (**c**) Optimal computation time

(a) Very low κ (i.e., $\kappa \leq 0.004$): Designed for solely energy minimization. In this region, all UE runs their CPU at the lowest cycle frequency f_n^{min}, thus T_{cmp}^* is determined by the last UEs that finish their computation with their minimum frequency.

(b) Low κ (i.e., $0.004 \leq \kappa \leq 0.1$): Designed for prioritized energy minimization. This region contains UEs of both $\mathcal{N}_2$ and $\mathcal{N}_3$. T_{cmp}^* is governed by which subset has a higher virtual computation deadline, which also determines the optimal CPU-cycle frequency of $\mathcal{N}_3$. Other UEs with light-loaded data, if exist, can run at the most energy-saving mode f_n^{min} yet still finish their task before T_{cmp}^* (i.e., $\mathcal{N}_2$).

(c) Medium κ (i.e., $0.1 \leq \kappa \leq 1$): Designed for balancing computation time and energy minimization. All UEs belong to $\mathcal{N}_3$ with their optimal CPU-cycle frequency strictly inside the feasible set.

(d) High κ (i.e., $\kappa \geq 1$): Designed for prioritized computation time minimization. High value κ can ensure the existence of $\mathcal{N}_1$, consisting the most "bottleneck" UEs (i.e., heavy-loaded data and/or low f_n^{max}) that runs their maximum CPU-cycle in (3.30) (top) and thus determines the optimal computation time T_{cmp}^*. The other "non-bottleneck" UEs either (i) adjust a "right" CPU-cycle to save the energy yet still maintain their computing time the same as T_{cmp}^* (i.e., $\mathcal{N}_3$), or (ii) can finish the computation with minimum frequency before the "bottleneck" UEs (i.e., $\mathcal{N}_2$) as in (3.30) (bottom).

SUB2 Solution

Before characterizing the solution to **SUB2**, from (3.12) and (3.24), we first define
two bounded values for τ_n as follows

$$\tau_n^{max} = \frac{s_n}{B \ln(h_n N_0^{-1} p_n^{min} + 1)}, \tag{3.31}$$

$$\tau_n^{min} = \frac{s_n}{B \ln(h_n N_0^{-1} p_n^{max} + 1)}, \tag{3.32}$$

which are the maximum and minimum possible fractions of T_{com} that UE n can
achieve by transmitting with its minimum and maximum power, respectively. We
also define a new function $g_n : \mathbb{R} \rightarrow \mathbb{R}$ as

$$g_n(\kappa) = \frac{s_n/B}{1 + W\left(\frac{\kappa N_0^{-1} h_n - 1}{e}\right)}, \tag{3.33}$$

where $W(.)$ is the Lambert W-function. We can consider $g_n(.)$ as an indirect "power
control" function that helps UE n control the amount of time it should transmit an
amount of data s_n by adjusting the power based on the weight κ. This function is
strictly decreasing (thus its inverse function $g_n^{-1}(\cdot)$ exists) reflecting that when we
put more priority on minimizing the communication time (i.e., high κ), UE n should
raise the power to finish its transmission with less time (i.e., low τ_n).

Lemma 3.2 *The solution to* **SUB2** *is as follows*

(a) If $\kappa \le g_n^{-1}(\tau_n^{max})$, then

$$\tau_n^* = \tau_n^{max} \tag{3.34}$$

(b) If $g_n^{-1}(\tau_n^{max}) < \kappa < g_n^{-1}(\tau_n^{min})$, then

$$\tau_n^{min} < \tau_n^* = g_n(\kappa) < \tau_n^{max} \tag{3.35}$$

(c) If $\kappa \ge g_n^{-1}(\tau_n^{min})$, then

$$\tau_n^* = \tau_n^{min}, \tag{3.36}$$

and $T_{com}^* = \sum_{n=1}^{N} \tau_n^*$.

This lemma can be explained in the view of network economics as follows. If we
interpret the **FEDL** system as the buyer and UEs as sellers with the UE powers as
commodities, then the inverse function $g_n^{-1}(\cdot)$ is interpreted as the price of energy
that UE n is willing to accept by providing power service for **FEDL** to reduce the
learning time. There are two properties of this function: (i) the price increases with

respect to UE power, and (ii) the price sensitivity depends on UEs characteristics, e.g., UEs with better channel quality can have a lower price, whereas UEs with larger data size s_n will have a higher price. Thus, each UE n will compare its energy price $g_n^{-1}(\cdot)$ with the "offer" price κ by the system to decide how much power it is willing to "sell". Then, there are three cases corresponding to the solutions to SUB2.

(a) Low offer: If the offer price κ is lower than the minimum price request $g_n^{-1}(\tau_n^{max})$, UE n will sell its lowest service by transmitting with the minimum power p_n^{min}.
(b) Medium offer: If the offer price κ is within the range of an acceptable price range, UE n will find a power level such that the corresponding energy price will match the offer price.
(c) High offer: If the offer price κ is higher than the maximum price request $g_n^{-1}(\tau_n^{min})$, UE n will sell its highest service by transmitting with the maximum power p_n^{max}.

Lemma 3.2 is further illustrated in Fig. 3.3, showing how the solution to SUB2 varies with respect to κ. It is observed from this figure that due to the UE heterogeneity of channel gain, $\kappa = 0.1$ is a medium offer to UEs 2, 3, and 4, but a high offer to UE 1, and a low offer to UE 5.

While SUB1 and SUB2 solutions share the same threshold-based dependence, we observe their differences as follows. In SUB1 solution, the optimal CPU-cycle frequency of UE n depends on the optimal T_{cmp}^*, which in turn depends on the loads (i.e., $\frac{c_n D_n}{f_n}$, $\forall n$) of all UEs. Thus all UE load information is required for the computation phase. On the other hand, in SUB2 solution, each UE n can independently choose its optimal power by comparing its price function $g_n^{-1}(\cdot)$ with κ so that collecting UE information is not needed. The reason is that the synchronization of computation time in constraint (3.20) of SUB1 requires all UE

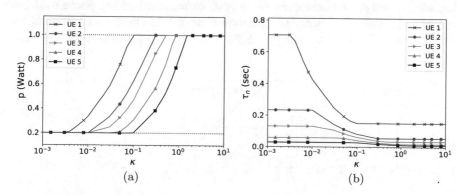

Fig. 3.3 The solution to SUB2 with five UEs. The numerical setting is the same as that of Fig. 3.2. (**a**) UEs' optimal transmission power (**b**) UEs' optimal transmission time

loads, whereas the UEs' time-sharing constraint (3.23) of **SUB2** can be decoupled by comparing with the fixed "offer" price κ.

SUB3 Solution

We observe that the solutions to **SUB1** and **SUB2** have no dependence on θ so that the optimal T_{com}^*, T_{cmp}^*, f^*, and τ^* can be determined based on κ according to Lemmas 3.1 and 3.2. However, these solutions will affect the third sub-problem of FEDL, as will be shown in what follows.

SUB3:

min. $K(\theta)\big[E_{glob}(f^*, \tau^*, \theta) + \kappa\, T_{glob}(T_{cmp}^*, T_{com}^*, \theta)\big]$

s.t. $0 \le \theta \le 1.$ (3.37)

Lemma 3.3 *There exists a unique solution θ^* of the convex problem* **SUB3** *satisfying the following equation:*

$$\frac{1}{\eta} = \log\!\big(e^{1/\theta^*}\theta^*\big) \tag{3.38}$$

where

$$\eta = \frac{\sum_{n=1}^{N} E_n^{cmp}(f_n^*) + \kappa T_{cmp}^*}{\sum_{n=1}^{N}\big[E_n^{cmp}(f_n^*) + E_n^{com}(\tau_n^*)\big] + \kappa\big[T_{cmp}^* + T_{com}^*\big]}. \tag{3.39}$$

The convexity and unique solution to **SUB3** are illustrated in Fig. 3.4a and b, respectively. From its definition, η is the fraction of total computation cost (including all UE energy and computing time cost) of the computation phase over the aggregated communication and computation costs. We then have some observations from Lemma 3.3. First, according to (3.38), it can be shown that $0 < \theta^* < \eta < 1$,

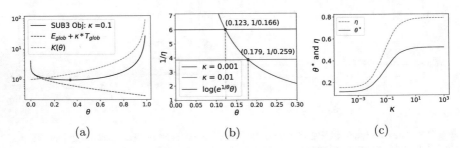

(a)   (b)   (c)

Fig. 3.4 The solution to **SUB3** with five UEs: (**a**) Convexity of **SUB3**, (**b**) Unique θ^* for each η, and (**c**) Impact of κ on η and θ^*. The numerical setting is the same as that of Fig. 3.2

which is also illustrated in Fig. 3.4b (with two different values of κ) and Fig. 3.4c. Thus, small computation cost (compared to communication cost) implies small θ^*, which means UEs need to run a large number of local iterations in the computation phase to reduce the number of global iterations $K(\theta^*)$ due to more expensive communication cost. Second, the impact of κ on η and θ^* is illustrated in Fig. 3.4c as follows.

(a) When κ is small enough such that the energy cost dominates the time cost, e.g., $\kappa \leq 10^{-3}$, then $\eta \approx \dfrac{\sum_{n=1}^{N} E_n^{cmp}(f_n^*)}{\sum_{n=1}^{N}\left[E_n^{cmp}(f_n^*)+E_n^{com}(\tau_n^*)\right]}$ (which approaches to a constant when κ falls into case a) of Corollary 3.1 and Lemma 3.2). Small values of η in this case (i.e, $\eta \leq 0.17$) indicate that computation energy is much smaller than communication energy; thus UEs are better to perform more local computations and less communications, explaining the corresponding small value of θ^* (i.e., $\theta^* \leq 0.12$).

(b) When κ is in a range such that the energy cost is comparable to the time cost, e.g., $10^{-3} \leq \kappa \leq 10$, increasing κ causes an increase η (because $\eta < 1$), indicating that the computation cost is increasingly more expensive than communication cost. Thus θ^* also increases, reflecting **FEDL** preference on more communication and less computation.

(c) When κ is large enough such that the energy cost dominates the time cost (i.e., $\kappa \geq 10$), then $\eta \approx \dfrac{T_{cmp}^*}{T_{cmp}^*+T_{com}^*}$ (which approaches to a constant when κ falls into case d) of Corollary 3.1 and case (c) of Lemma 3.2). Large values of η in this case (i.e., $\eta \geq 0.79$) indicate that computation time is much larger than communication time; thus UEs are better to perform less local computations and more communications, explaining the corresponding large value of θ^* (i.e., $\theta^* \geq 0.53$).

FEDL Solution

Theorem 3.1 *The globally optimal solution to* **FEDL** *is the combined solutions to three sub-problems* **SUB1**, **SUB2**, *and* **SUB3**.

The proof of this theorem is straightforward. The idea is to use the KKT condition to find the stationary points of **FEDL** . Then we can decompose the KKT condition equations into three groups, each of them matches exactly to the KKT condition of **SUB1**, **SUB2**, and **SUB3**, which can be solved for unique closed-form solution as in Lemmas 3.1, 3.2, and 3.3, respectively. Thus this unique stationary point is also the globally optimal solution to **FEDL**.

We then have some discussions on the combined solution to **FEDL**. First, we see that **SUB1** and **SUB2** solutions can be characterized independently, which can be explained that each UE often has two separate processors: one CPU for mobile applications and another baseband processor for the radio control function. Second, neither **SUB1** nor **SUB2** depends on θ because the communication phase in SUB2 is clearly not affected by the local accuracy of the computing problem, whereas SUB2 considers the computation cost in one local iteration. However, the solutions to SUB1 and SUB2, which can reveal how much communication cost is more expensive than computation cost, are decisive factors to determine the optimal level of local accuracy. Therefore, we can sequentially solve SUB1 and SUB2 first, then SUB3 to achieve the optimal solutions to FEDL.

3.2.4   Numerical Results

In this subsection, both the communication and computation models follow the same setting as in Fig. 3.2, except the number of UEs is increased to 50, and all UEs have the same $f_n^{max} = 2.0\,\mathrm{GHz}$, $c_n = 20$ cycles/bit. Furthermore, we define two new parameters, addressing the UE heterogeneity regarding computation and communication phases in FEDL, respectively, as follows

$$L_{cmp} = \frac{\max_{n \in \mathcal{N}} \frac{c_n D_n}{f_n^{max}}}{\min_{n \in \mathcal{N}} \frac{c_n D_n}{f_n^{min}}} \tag{3.40}$$

$$L_{com} = \frac{\max_{n \in \mathcal{N}} \tau_n^{min}}{\min_{n \in \mathcal{N}} \tau_n^{max}}. \tag{3.41}$$

We see that higher values of L_{cmp} and L_{com} indicate higher levels of UE heterogeneity. For example, $L_{cmp} = 1$ ($L_{com} = 1$) can be considered as a high heterogeneity level due to unbalanced data distributed and/or UE configuration (unbalanced channel gain distribution) such that UE with their minimum frequency (maximum transmission power) still have the same computation (communication) time as those with maximum frequency (minimum transmission power). The level of heterogeneity is controlled by two different settings. To vary L_{cmp}, the training size D_n is generated with the fraction $\frac{D^{min}}{D^{max}} \in \{1., 0.2, 0.001\}$ but the average data of all UEs is kept at the same value 7.5 MB for varying values of L_{cmp}. On the other hand, to vary L_{com}, the distance between these devices and the BS is generated such that $\frac{d^{min}}{d^{max}} \in \{1., 0.2, 0.001\}$ but the average distance of all UEs is maintained at 26 m for different values of L_{com}. Here D^{min} and D^{max} (d^{min} and d^{max}) are minimum and maximum data size (BS-to-UE distance), respectively. In all scenarios, we fix $L_{cmp} = 0.3$ when varying L_{com} and fix $L_{com} = 0.48$ when varying L_{cmp}.

Impact of UE Heterogeneity

We first examine the impact of UE heterogeneity on SUB1 and SUB2 in Fig. 3.5, which shows that increasing L_{cmp} and L_{com} enforces the optimal f_n^* and τ_n^* having more diverse values and thus makes increase the computation and communication

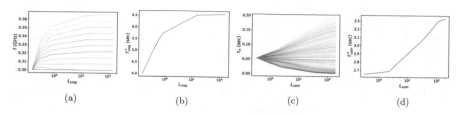

(a) (b) (c) (d)

Fig. 3.5 Impact of UE heterogeneity on SUB1 and SUB2 with $\kappa = 0.07$. (**a**) Impact of L_{cmp} on f_n^*. (**b**) Impact of L_{cmp} on T_{cmp}^*. (**c**) Impact of L_{com} on τ_n^*. (**d**) Impact of L_{com} on T_{com}^*

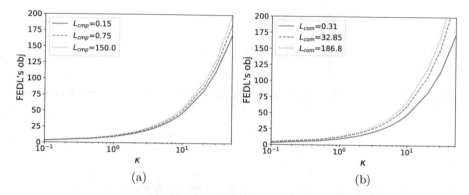

(a) (b)

Fig. 3.6 Impact of UE heterogeneity on FEDL. (**a**) Impact of L_{cmp}. (**b**) Impact of L_{com}

time T_{cmp}^* and T_{com}^*, respectively. As expected, we observe that the high level of UE heterogeneity has a negative impact on the FEDL system, as illustrated in Fig. 3.6a and b, such that the total cost (objective of FEDL) is increased with a higher value of L_{cmp} and L_{com} respectively. However, in this setting, when T_{cmp} is comparable to T_{com}, e.g., 6.2 versus 2.9 s at $L_{cmp} = L_{com} = 10$, the impact of L_{com} on the total cost is more profound than that of L_{cmp}, e.g., at $\kappa = (0.1, 1, 10)$, the total cost of FEDL increases (1.09, 1.11, 1.10) times and (1.62, 1.40, 1.43) times, when L_{cmp} and L_{com} are increased from 0.15 to 150, and from 0.31 to 186.8, respectively.

On the other hand, with a different setting such that T_{cmp} dominates T_{com}, e.g., 80 versus 7.8 s at $L_{cmp} = L_{com} = 10$, the impacts of L_{cmp} and L_{com} on total cost are comparable, e.g., at $\kappa = (0.1, 1, 10)$, the total cost of FEDL increases (1.14, 1.72, 1.65) times and (1.36, 1.21, 1.23) times, when L_{cmp} and L_{com} are increased from 0.05 to 50, and from 0.17 to 181.42, respectively.

Pareto Optimal Trade-off

We next illustrate the Pareto curve in Fig. 3.7. This curve shows the trade-off between the conflicting goals of minimizing the time cost $K(\theta)T_{glob}$ and energy cost $K(\theta)E_{glob}$, in which we can decrease one type of cost yet with the expense of increasing the other one. This figure also shows that the Pareto curve of FEDL is

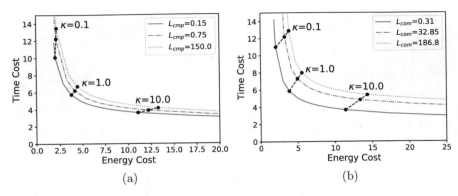

Fig. 3.7 Pareto-optimal points of FEDL. (**a**) Impact of L_{cmp} and κ. (**b**) Impact of L_{com} and κ

more efficient when the system has a low level of UE heterogeneity (i.e., small L_{cmp} and/or L_{com}).

Impact of η

We quantify the impact of η on the optimal θ^* by varying κ in Fig. 3.8. Similar to the observations after Lemma 3.3, when κ is very small or very large, the value of η, which drives the corresponding value of θ^*, is determined by the proportions shown in Fig. 3.9a or 3.9b, respectively, which also drives the corresponding value of θ^*. However, in this setting with 50 UEs, when κ is very large, η and θ^* decrease to small values, which is in contrast to the scenario with 5 UEs in Fig. 3.4c (i.e, large η and θ^* with large κ). The main reason for this difference is that communication time scales with the increasing number of UEs due to wireless sharing nature, which makes the time portion small when κ is large, as shown in Fig. 3.9b. The final observation from Figs. 3.8 and 3.9 is that the higher L_{cmp} (L_{com}), the higher T^*_{cmp}

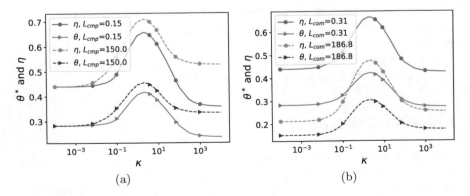

Fig. 3.8 Impact of κ on η and θ^*

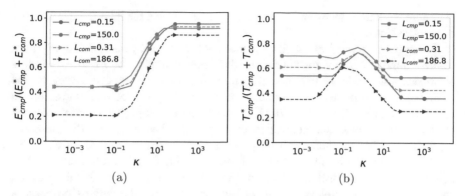

Fig. 3.9 Proportion of computation energy and time

(T_{com}^*), and thus higher (lower) portion of computation time, which makes higher (lower) values of η and θ^*.

3.3 Wireless Federated Learning: Resource Allocation and Transmit Power Allocation

In this section, we introduce the optimization of resource allocation and transmit power for FL implemented over wireless networks. In particular, we first provide a detailed literature review on the optimization of resource management for FL. Then, we introduce a representative work that jointly optimize wireless resources such as transmit power, resource block (RB) allocation and user selection for FL implemented over wireless networks.

3.3.1 Motivation

Standard machine learning approaches require devices to transmit their collected training data to a parameter sever for training purpose [66]. However, due to privacy and limited communication resources, it is impractical for all users that participate in learning to transmit all of their collected data to a parameter sever. This, in turn, motivates the development of distributed learning frameworks that allow devices to use individually collected data to train a learning model locally. One of the most promising of such distributed learning frameworks is FL developed in [67]. FL enables users to collaboratively learn a shared machine learning model while keeping their collected data on their devices [3, 68–70]. However, to train an FL algorithm in a distributed manner, the users must transmit the trained parameters over wireless links which can introduce training errors and the inherent unreliability of wireless links.

Recently, a number of existing works such as in [67] and[18, 71–78] have studied important problems related to the implementation of FL over wireless networks. The works in [67] and [71] provided a comprehensive survey on the design of FL algorithms and introduced various challenges, problems, and solutions for enhancing FL effectiveness. In [72], the authors developed two update methods to reduce the uplink communication costs for FL. The work in [18] presented a practical update method for a deep FL algorithm and conducted an extensive empirical evaluation for five different FL models using four datasets. An echo state network-based FL algorithm is developed in [73] to analyze and predict the location and orientation for wireless virtual reality users. In [74], the authors proposed a novel FL algorithm that can minimize the communication cost. The authors in [75] studied the problem of joint power and resource allocation for ultra-reliable low latency communication in vehicular networks. The work in [76] developed a new approach to minimize the computing and transmission delay for FL algorithms. While interesting, these prior works [67] and [18, 71–76] assumed that wireless networks can readily integrate FL algorithms. However, in practice, due to the unreliability of the wireless channels and to the wireless resource limitations (e.g., in terms of bandwidth and power), FL algorithms will encounter training errors due to the wireless links [77]. For example, symbol errors introduced by the unreliable nature of the wireless channel and by resource limitations can impact the quality and correctness of the FL updates among users. Such errors will, in turn, affect the performance of FL algorithms, as well as their convergence speed. Moreover, due to the wireless bandwidth limitations, the number of users that can perform FL is limited; a design issue that is ignored in [67] and [18, 71–76]. Furthermore, due to limited energy consumption of each user's device and strict delay requirement of FL, not all wireless users can perform FL algorithms. Therefore, one must select the appropriate users to perform FL algorithms and optimize the performance of FL. In practice, to effectively deploy FL over real-world wireless networks, it is necessary to investigate how the wireless factors affect the performance of FL algorithms. Here, we note that, although some works such as [68] and [31, 77–83] have studied communication aspects of FL, these works are limited in several ways. First, the works in [68, 77], and [79] only provided a high-level exposition of the challenges of communication in FL. Meanwhile, the authors in [31, 78–82] did not consider the effect of packet transmission errors on the performance of FL. The authors in [83] developed an analytical model to characterize the effect of packet transmission errors on the FL performance. However, the work in [83] only measured the effectiveness of three different scheduling policies and, hence, did not find optimal user selection and RB allocation to optimize the FL performance.

Next, we introduce a representative work that jointly optimize wireless resources such as transmit power, RB allocation and user selection to minimize the training loss of FL.

3.3.2 System Model

Consider a cellular network in which one BS and a set $\mathcal{U}$ of U users cooperatively perform an FL algorithm for data analysis and inference. For example, the network can execute an FL algorithm to sense the wireless environment and generate a holistic radio environment mapping. The use of FL for such applications is important because the data related to the wireless environment is distributed across the network [3] and the BS cannot collect all of this scattered data to implement a centralized learning algorithm. FL enables the BS and the users to collaboratively learn a shared learning model while keeping all of the training data at the device of each user. In an FL algorithm, each user will use its collected training data to train an FL model. For example, for radio environment mapping, each user will collect the data related to the wireless environment for training an FL model. Hereinafter, the FL model that is trained at the device of each user (using the data collected by the user itself) is called the *local FL model*. The BS is used to integrate the local FL models and generate a shared FL model. This shared FL model is used to improve the local FL model of each user so as to enable the users to collaboratively perform a learning task without training data transfer. Hereinafter, the FL model that is generated by the BS using the local FL models of its associated users is called the *global FL model*. As shown in Fig. 3.10, the *uplink* from the users to the BS is used to transmit the local FL model parameters while the *downlink* is used to transmit the global FL model parameters.

Machine Learning Model

In our model, each user i collects a matrix $X_i = \left[x_{i1}, \ldots, x_{iK_i}\right]$ of input data, where K_i is the number of the samples collected by each user i and each element

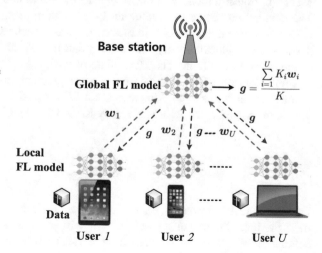

Fig. 3.10 The architecture of an FL algorithm that is being executed over a wireless network with multiple devices and a single base station

x_{ik} is an input vector of the FL algorithm. The size of x_{ik} depends on the specific FL task. Our approach, however, is applicable to any generic FL algorithm and task. Let y_{ik} be the output of x_{ik}. For simplicity, we consider an FL algorithm with a single output, however, our approach can be readily generalized to a case with multiple outputs [72]. The output data vector for training the FL algorithm of user i is $y_i = [y_{i1}, \ldots, y_{iK_i}]$. We define a vector w_i to capture the parameters related to the local FL model that is trained by X_i and y_i. In particular, w_i determines the local FL model of each user i. For example, in a linear regression learning algorithm, $x_{ik}^T w_i$ represents the predicted output and w_i is a weight vector that determines the performance of the linear regression learning algorithm. For each user i, the local training problem seeks to find the optimal learning model parameters w_i^* that minimize its training loss. The training process of an FL algorithm is done in a way to solve the following optimization problem:

$$\min_{w_1, \ldots, w_U} \frac{1}{K} \sum_{i=1}^{U} \sum_{k=1}^{K_i} f(w_i, x_{ik}, y_{ik}), \qquad (3.42)$$

$$\text{s. t. } w_1 = w_2 = \ldots = w_U = g, \qquad \forall i \in \mathcal{U}, \qquad (3.42a)$$

where $K = \sum_{i=1}^{U} K_i$ is total size of training data of all users and g is the global FL model that is generated by the BS and $f(w_i, x_{ik}, y_{ik})$ is a loss function. The loss function captures the performance of the FL algorithm. Constraint (3.42a) is used to ensure that, once the FL algorithm converges, all of the users and the BS will share the same FL model for their learning task. This captures the fact that the purpose of an FL algorithm is to enable the users and the BS to learn an optimal global FL model without data transfer. To solve (3.42), the BS will transmit the parameters g of the global FL model to its users so that they train their local FL models. Then, the users will transmit their local FL models to the BS to update the global FL model. The detailed procedure of training an FL algorithm to minimize the loss function in (3.42) is shown in Fig. 3.11. In FL, the update of each user i's local FL model w_i depends on the global model g while the update of the global model g depends on all of the users' local FL models. The update of the local FL model w_i depends on the learning algorithm. For example, one can use gradient descent, stochastic gradient descent, or randomized coordinate descent [72] to update the local FL model. The update of the global model g is given by Konečný et al. [72]

$$g_t = \frac{\sum_{i=1}^{U} K_i w_{i,t}}{K}. \qquad (3.43)$$

During the training process, each user will first use its training data X_i and y_i to train the local FL model w_i and then, it will transmit w_i to the BS via wireless cellular links. Once the BS receives the local FL models from all participating

Fig. 3.11 The learning procedure of an FL algorithm

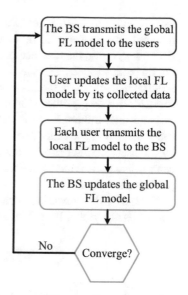

users, it will update the global FL model based on (3.43) and transmit the global FL model g to all users to optimize the local FL models. As time elapses, the BS and users can find their optimal FL models and use them to minimize the loss function in (3.42). Since all of the local FL models are transmitted over wireless cellular links, once they are received by the BS, they may contain erroneous symbols due to the unreliable nature of the wireless channel, which, in turn, will have a significant impact on the performance of FL. Meanwhile, the BS must update the global FL model once it receives all of the local FL models from its users and, hence, the wireless transmission delay will significantly affect the convergence of the FL algorithm. In consequence, to deploy FL over a wireless network, *one must jointly consider the wireless and learning performance and factors.*

Transmission Model

For uplink, we assume that an orthogonal frequency division multiple access (OFDMA) technique in which each user occupies one RB. The uplink rate of user i transmitting its local FL model parameters to the BS is given by

$$c_i^U (r_i, P_i) = \sum_{n=1}^{R} r_{i,n} B^U \mathbb{E}_{h_i} \left(\log_2 \left(1 + \frac{P_i h_i}{I_n + B^U N_0} \right) \right), \tag{3.44}$$

where $r_i = [r_{i,1}, \ldots, r_{i,R}]$ is an RB allocation vector with R being the total number of RBs, $r_{i,n} \in \{0, 1\}$ and $\sum_{n=1}^{R} r_{i,n} = 1$; $r_{i,n} = 1$ indicates that RB n is allocated to user i, and $r_{i,n} = 0$, otherwise; B^U is the bandwidth of each RB and P_i is the transmit power of user i; $h_i = o_i d_i^{-2}$ is the channel gain between user i and the BS with d_i being the distance between user i and the BS and o_i being the Rayleigh fading parameter; $\mathbb{E}_{h_i}(\cdot)$ is the expectation with respect to h_i; N_0 is the noise power spectral density; I_n is the interference caused by the users that are located in other service areas (e.g., other BSs not participating in the FL algorithm) and use RB n. Note that, although we ignore the optimization of resource allocation for the users located at the other service areas, we must consider the interference caused by the users in other service areas (if they are sharing RBs with the considered FL users), since this interference may significantly affect the packet error rates and the performance of FL.

Similarly, the downlink data rate achieved by the BS when transmitting the parameters of global FL model to each user i is given by

$$c_i^D = B^D \mathbb{E}_{h_i} \left(\log_2 \left(1 + \frac{P_B h_i}{I^D + B^D N_0} \right) \right), \tag{3.45}$$

where B^D is the bandwidth that the BS used to broadcast the global FL model of each user i; P_B is the transmit power of the BS; I^D is the interference caused by other BSs not participating in the FL algorithm. Given the uplink data rate c_i^U in (3.44) and the downlink data rate c_i^D in (3.45), the transmission delays between user i and the BS over uplink and downlink are respectively specified as

$$l_i^U(r_i, P_i) = \frac{Z(w_i)}{c_i^U(r_i, P_i)}, \tag{3.46}$$

$$l_i^D = \frac{Z(g)}{c_i^D}, \tag{3.47}$$

where function $Z(x)$ is the data size of x which is defined as the number of bits that the users or the BS require to transmit vector x over wireless links. In particular, $Z(w_i)$ represents the number of bits that each user i requires to transmit local FL model w_i to the BS while $Z(g)$ is the number of bits that the BS requires to transmit the global FL model g to each user. Here, $Z(w_i)$ and $Z(g)$ are determined by the type of implemented FL algorithm. From (3.43), we see that the number of elements in the global FL model g is similar to that of each user i's local FL model w_i. Hence, we assume $Z(w_i) = Z(g)$.

Packet Error Rates

For simplicity, we assume that each local FL model w_i will be transmitted as a single packet in the uplink. A cyclic redundancy check (CRC) mechanism is used to check the data errors in the received local FL models at the BS. In particular, $C(w_i) = 0$

indicates that the local FL model received by the BS contains data errors; otherwise, we have $C(\boldsymbol{w}_i) = 1$. The packet error rate experienced by the transmission of each local FL model $\boldsymbol{w}_i$ to the BS is given by Xi et al. [84]

$$q_i(\boldsymbol{r}_i, P_i) = \sum_{n=1}^{R} r_{i,n} q_{i,n},$$

(3.48)

where $q_{i,n} = \mathbb{E}_{h_i}\left(1 - \exp\left(-\frac{m(I_n + B^{\mathrm{U}} N_0)}{P_i h_i}\right)\right)$ is the packet error rate over RB n with m being a waterfall threshold [84].

In the considered system, whenever the received local FL model contains errors, the BS will not use it for the update of the global FL model. We also assume that the BS will not ask the corresponding users to resend their local FL models when the received local FL models contain data errors. Instead, the BS will directly use the remaining correct local FL models to update the global FL model. As a result, the global FL model in (3.43) can be written as

$$g(\boldsymbol{a}, \boldsymbol{P}, \boldsymbol{R}) = \frac{\sum_{i=1}^{U} K_i a_i \boldsymbol{w}_i C(\boldsymbol{w}_i)}{\sum_{i=1}^{U} K_i a_i C(\boldsymbol{w}_i)},$$

(3.49)

where

$$C(\boldsymbol{w}_i) = \begin{cases} 1, & \text{with probability } 1 - q_i(\boldsymbol{r}_i, P_i), \\ 0, & \text{with probability } q_i(\boldsymbol{r}_i, P_i), \end{cases}$$

(3.50)

$\boldsymbol{a} = [a_1, \ldots, a_U]$ is the vector of the user selection index with $a_i = 1$ indicating that user i performs the FL algorithm and $a_i = 0$, otherwise, $\boldsymbol{R} = [\boldsymbol{r}_1, \cdots, \boldsymbol{r}_U]$, $\boldsymbol{P} = [P_1, \cdots, P_U]$, $\sum_{i=1}^{U} K_i a_i C(\boldsymbol{w}_i)$ is the total number of training data samples, which depends on the user selection vector $\boldsymbol{a}$ and packet transmission $C(\boldsymbol{w}_i)$, $K_i a_i \boldsymbol{w}_i C(\boldsymbol{w}_i) = 0$ indicates that the local FL model of user i contains data errors and, hence, the BS will not use it to generate the global FL model, and $g(\boldsymbol{a}, \boldsymbol{P}, \boldsymbol{R})$ is the global FL model that explicitly incorporates the effect of wireless transmission. From (3.49), we see that the global FL model also depends on the resource allocation matrix $\boldsymbol{R}$, user selection vector $\boldsymbol{a}$, and transmit power vector $\boldsymbol{P}$.

Energy Consumption Model

In our network, the energy consumption of each user consists of the energy needed for two purposes: (a) Transmission of the local FL model and (b) Training of the local FL model. The energy consumption of each user i is given by Pan et al. [85]

$$e_i(\boldsymbol{r}_i, P_i) = \varsigma \omega_i \vartheta^2 Z(\boldsymbol{w}_i) + P_i l_i^{\mathrm{U}}(\boldsymbol{r}_i, P_i), \tag{3.51}$$

where ϑ is the frequency of the central processing unit (CPU) clock of each user i, ω_i is the number of CPU cycles required for computing per bit data of user i, which is assumed to be equal for all users, and ς is the energy consumption coefficient depending on the chip of each user i's device. In (3.51), $\varsigma \omega_i \vartheta^2 Z(\boldsymbol{w}_i)$ is the energy consumption of user i training the local FL model at its own device and $P_i l_i^{\mathrm{U}}(\boldsymbol{r}_i, P_i)$ represents the energy consumption of local FL model transmission from user i to the BS. Note that, since the BS can have continuous power supply, we do not consider the energy consumption of the BS in our optimization problem.

Problem Formulation

To jointly design the wireless network and the FL algorithm, we now formulate an optimization problem whose goal is to minimize the training loss, while factoring in the wireless network parameters. This minimization problem includes optimizing transmit power allocation as well as resource allocation for each user. The minimization problem is given by

$$\min_{\boldsymbol{a}, \boldsymbol{P}, \boldsymbol{R}} \frac{1}{K} \sum_{i=1}^{U} \sum_{k=1}^{K_i} f(g(\boldsymbol{a}, \boldsymbol{P}, \boldsymbol{R}), x_{ik}, y_{ik}), \tag{3.52}$$

$$\text{s. t. } a_i, r_{i,n} \in \{0, 1\}, \quad \forall i \in \mathcal{U}, n = 1, \ldots, R, \tag{3.52a}$$

$$\sum_{n=1}^{R} r_{i,n} = a_i, \quad \forall i \in \mathcal{U}, \tag{3.52b}$$

$$l_i^{\mathrm{U}}(\boldsymbol{r}_i, P_i) + l_i^{\mathrm{D}} \leq \gamma_{\mathrm{T}}, \quad \forall i \in \mathcal{U}, \tag{3.52c}$$

$$e_i(\boldsymbol{r}_i, P_i) \leq \gamma_{\mathrm{E}}, \quad \forall i \in \mathcal{U}, \tag{3.52d}$$

$$\sum_{i \in \mathcal{U}} r_{i,n} \leq 1, \forall n = 1, \ldots, R, \tag{3.52e}$$

$$0 \leq P_i \leq P_{\max}, \quad \forall i \in \mathcal{U}, \tag{3.52f}$$

where γ_{T} is the delay requirement for implementing the FL algorithm, γ_{E} is the energy consumption of the FL algorithm, and B is the total downlink bandwidth. (3.52a) and (3.52b) indicates that each user can occupy only one RB for uplink data transmission. (3.52c) is the delay needed to execute the FL algorithm at each learning step. (3.52d) is the energy consumption requirement of performing an FL algorithm at each learning step. (3.52e) indicates that each uplink RB can be allocated to at most one user. (3.52f) is a maximum transmit power constraint.

From (3.52), we can see that the user selection vector a, the RB allocation matrix R, and the transmit power vector P will not change during the FL training process and the optimized a, R, and P must meet the delay and energy consumption requirements at each learning step in (3.52c) and (3.52d).

From (3.48) and (3.49), we see that the transmit power and resource allocation determine the packet error rate, thus affecting the update of the global FL model. In consequence, the loss function of the FL algorithm in (3.52) depends on the resource allocation and transmit power. Moreover, (3.52c) shows that, in order to perform an FL algorithm, the users must satisfy a specific delay requirement. In particular, in an FL algorithm, the BS must wait to receive the local model of each user before updating its global FL model. Hence, transmission delay plays a key role in the FL performance. In a practical FL algorithm, it is desirable that all users transmit their local FL models to the BS simultaneously. From (3.52d), we see that to perform the FL algorithm, a given user must have enough energy to transmit and update the local FL model throughout the FL iterative process. If this given user does not have enough energy, the BS should choose this user to participate in the FL process. In consequence, in order to implement an FL algorithm in a real-world network, the wireless network must provide low energy consumption and latency, and highly reliable data transmission.

3.3.3 Convergence Analysis

To solve (3.52), we first need to analyze how the packet error rate affects the performance of the FL. To find the relationship between the packet error rates and the FL performance, we must first analyze the convergence rate of FL. However, since the update of the global FL model depends on the instantaneous signal-to-interference-plus-noise ratio (SINR), we can analyze only the expected convergence rate of FL. Here, we first analyze the expected convergence rate of FL. Then, we show how the packet error rate affects the performance of the FL in (3.52).

In the studied network, the users adopt a standard gradient descent method to update their local FL models as done in [72]. Therefore, during the training process, the local FL model w_i of each selected user i ($a_i = 1$) at step t is

$$
w_{i,t+1} = g_t (a, P, R) - \frac{\lambda}{K_i} \sum_{k=1}^{K_i} \nabla f \left(g_t (a, P, R), x_{ik}, y_{ik} \right), \tag{3.53}
$$

where λ is the learning rate and $\nabla f \left(g_t (a, P, R), x_{ik}, y_{ik} \right)$ is the gradient of $f \left(g_t (a, P, R), x_{ik}, y_{ik} \right)$ with respect to $g_t (a, P, R)$.

We assume that $F\left(\boldsymbol{g}\right)=\frac{1}{K}\sum_{i=1}^{U}\sum_{k=1}^{K_i}f\left(\boldsymbol{g},\boldsymbol{x}_{ik},y_{ik}\right)$ and $F_i\left(\boldsymbol{g}\right)=\sum_{k=1}^{K_i}f\left(\boldsymbol{g},\boldsymbol{x}_{ik},y_{ik}\right)$ where $\boldsymbol{g}$ is short for $\boldsymbol{g}\left(\boldsymbol{a},\boldsymbol{P},\boldsymbol{R}\right)$. Based on (3.53), the update of global FL model $\boldsymbol{g}$ at step t is given by

$$\boldsymbol{g}_{t+1} = \boldsymbol{g}_t - \lambda\left(\nabla F\left(\boldsymbol{g}_t\right) - \boldsymbol{o}\right), \tag{3.54}$$

where $\boldsymbol{o} = \nabla F\left(\boldsymbol{g}_t\right) - \dfrac{\sum_{i=1}^{U}a_i\sum_{k=1}^{K_i}\nabla f(\boldsymbol{g},\boldsymbol{x}_{ik},y_{ik})C(\boldsymbol{w}_i)}{\sum_{i=1}^{U}K_ia_iC(\boldsymbol{w}_i)}$. We also assume that the FL

algorithm converges to an optimal global FL model $\boldsymbol{g}^*$ after the learning steps. To derive the expected convergence rate of FL, we first make the following assumptions, as done in [70, 83].

- First, we assume that the gradient $\nabla F\left(\boldsymbol{g}\right)$ of $F\left(\boldsymbol{g}\right)$ is uniformly Lipschitz continuous with respect to $\boldsymbol{g}$ [86]. Hence, we have

$$\left\|\nabla F\left(\boldsymbol{g}_{t+1}\right) - \nabla F\left(\boldsymbol{g}_t\right)\right\| \leq L\|\boldsymbol{g}_{t+1} - \boldsymbol{g}_t\|, \tag{3.55}$$

  where L is a positive constant and $\|\boldsymbol{g}_{t+1} - \boldsymbol{g}_t\|$ is the norm of $\boldsymbol{g}_{t+1} - \boldsymbol{g}_t$.
- Second, we assume that $F\left(\boldsymbol{g}\right)$ is strongly convex with positive parameter μ, such that

$$F\left(\boldsymbol{g}_{t+1}\right) \geq F\left(\boldsymbol{g}_t\right) + \left(\boldsymbol{g}_{t+1} - \boldsymbol{g}_t\right)^T\nabla F\left(\boldsymbol{g}_t\right) + \frac{\mu}{2}\|\boldsymbol{g}_{t+1} - \boldsymbol{g}_t\|^2. \tag{3.56}$$

- We also assumed that $F\left(\boldsymbol{g}\right)$ is twice-continuously differentiable. Based on (3.55) and (3.56), we have

$$\mu\boldsymbol{I} \preceq \nabla^2 F\left(\boldsymbol{g}\right) \preceq L\boldsymbol{I}. \tag{3.57}$$

- We also assume that $\|\nabla f\left(\boldsymbol{g}_t,\boldsymbol{x}_{ik},y_{ik}\right)\|^2 \leq \zeta_1 + \zeta_2\|\nabla F\left(\boldsymbol{g}_t\right)\|^2$ with $\zeta_1, \zeta_2 \geq 0$.

These assumptions can be satisfied by several widely used loss functions such as the mean squared error, logistic regression, and cross entropy [86]. These popular loss functions can be used to capture the performance of implementing practical FL algorithms for identification, prediction, and classification. For future work, we can investigate how to extend our work for other non-convex loss functions. The expected convergence rate of the FL algorithms can now be obtained by the following theorem.

Theorem 3.2 *Given the transmit power vector $\boldsymbol{P}$, RB allocation matrix $\boldsymbol{R}$, user selection vector $\boldsymbol{a}$, optimal global FL model $\boldsymbol{g}^*$, and the learning rate $\lambda = \frac{1}{L}$, the upper bound of $\mathbb{E}\left(F\left(\boldsymbol{g}_{t+1}\right) - F\left(\boldsymbol{g}^*\right)\right)$ can be given by*

$$
\mathbb{E}\left(F\left(\boldsymbol{g}_{t+1}\right) - F\left(\boldsymbol{g}^*\right)\right) \le A^t \mathbb{E}\left(F\left(\boldsymbol{g}_0\right) - F\left(\boldsymbol{g}^*\right)\right)
$$

$$
+ \underbrace{\frac{2\zeta_1}{LK} \sum_{i=1}^{U} K_i \left(1 - a_i + a_i q_i\left(\boldsymbol{r}_i, P_i\right)\right) \frac{1 - A^t}{1 - A}}_{\text{Impact of wireless factors on FL convergence}}, \tag{3.58}
$$

where $A = 1 - \frac{\mu}{L} + \frac{4\mu\zeta_2}{LK} \sum_{i=1}^{U} K_i \left(1 - a_i + a_i q_i\left(\boldsymbol{r}_i, P_i\right)\right)$ and $\mathbb{E}\left(\cdot\right)$ is the expectation with respect to packet error rate.

Proof See in [29].

In Theorem 3.2, $\boldsymbol{g}_{t+1}$ is the global FL model that is generated based only on the local FL models of selected users ($a_i = 1$) at step $t + 1$. $\boldsymbol{g}^*$ is the optimal FL model that is generated based on the local FL models of all uses in an ideal setting with no wireless errors. From Theorem 3.2, we see that a gap, $\frac{2\zeta_1}{LK} \sum_{i=1}^{U} K_i \left(1 - a_i + a_i q_i\left(\boldsymbol{r}_i, P_i\right)\right) \frac{1 - A^t}{1 - A}$, exists between $\mathbb{E}\left(F\left(\boldsymbol{g}_t\right)\right)$ and $\mathbb{E}\left(F\left(\boldsymbol{g}^*\right)\right)$. This gap is caused by the packet errors and the user selection policy. As the packet error rate decreases, the gap between $\mathbb{E}\left(F\left(\boldsymbol{g}_t\right)\right)$ and $\mathbb{E}\left(F\left(\boldsymbol{g}^*\right)\right)$ decreases. Meanwhile, as the number of users that implement the FL algorithm increases, the gap also decreases. Moreover, as the packet error rate decreases, the value of A also decreases, which indicates that the convergence speed of the FL algorithm improves. Hence, it is necessary to optimize resource allocation, user selection, and transmit power for the implementation of any FL algorithm over a realistic wireless network. Theorem 3.2 can be extended to the case in which each local FL model needs to be transmitted over a large number of packets by replacing the packet error rate $q_i\left(\boldsymbol{r}_i, P_i\right)$ in (3.58) with the error rate of transmitting multiple packets to send the entire local FL model.

3.3.4 Optimization of RB Allocation and Transmit Power for FL Training Loss Minimization

In this subsection, we introduce how to solve the problem in (3.52). To solve the problem in (3.52), we must first simplify it. From Theorem 3.2, we can see that, to minimize the training loss in (3.52), we need to only minimize the gap, $\frac{2\zeta_1}{LK} \sum_{i=1}^{U} K_i \left(1 - a_i + a_i q_i\left(\boldsymbol{r}_i, P_i\right)\right) \frac{1 - A^t}{1 - A}$. When $A \ge 1$, the FL algorithm will not converge. In consequence, here, we only consider the minimization of the FL training loss when $A < 1$. Hence, as t is large enough, which captures the asymptotic convergence behavior of FL, we have $A^t = 0$. The gap can be rewritten as

$$\frac{2\zeta_1}{LK} \sum_{i=1}^{U} K_i \left(1 - a_i + a_i q_i \left(\mathbf{r}_i, P_i\right)\right) \frac{1 - A^t}{1 - A} =$$

$$\frac{\frac{2\zeta_1}{LK} \sum_{i=1}^{U} K_i \left(1 - a_i + a_i q_i \left(\mathbf{r}_i, P_i\right)\right)}{\frac{\mu}{L} - \frac{4\mu\zeta_2}{LK} \sum_{i=1}^{U} K_i \left(1 - a_i + a_i q_i \left(\mathbf{r}_i, P_i\right)\right)}. \tag{3.59}$$

From (3.59), we can observe that minimizing $\frac{2\zeta_1}{LK} \sum_{i=1}^{U} K_i \left(1 - a_i + a_i q_i \left(\mathbf{r}_i, P_i\right)\right)$ $\frac{1 - A^t}{1 - A}$ only requires minimizing $\sum_{i=1}^{U} K_i \left(1 - a_i + a_i q_i \left(\mathbf{r}_i, P_i\right)\right)$. Meanwhile, since $a_i = \sum_{n=1}^{R} r_{i,n}$ and $q_i \left(\mathbf{r}_i, P_i\right) = \sum_{n=1}^{R} r_{i,n} q_{i,n}$, we have $q_i \left(\mathbf{r}_i, P_i\right) \leq 1$, when $a_i = 1$, and $q_i \left(\mathbf{r}_i, P_i\right) = 0$, if $a_i = 0$. In consequence, we have $a_i q_i \left(\mathbf{r}_i, P_i\right) = q_i \left(\mathbf{r}_i, P_i\right)$. The problem in (3.52) can be simplified as

$$\min_{P,R} \sum_{i=1}^{U} K_i \left(1 - \sum_{n=1}^{R} r_{i,n} + q_i \left(\mathbf{r}_i, P_i\right)\right), \tag{3.60}$$

s. t. (3.52c)–(3.52f),

$$r_{i,n} \in \{0, 1\}, \quad \forall i \in \mathcal{U}, n = 1, \dots, R, \tag{3.60a}$$

$$\sum_{n=1}^{R} r_{i,n} \leq 1, \quad \forall i \in \mathcal{U}. \tag{3.60b}$$

Next, we first use RB allocation matrix $\mathbf{R}$ to represent the optimal transmit power for each user. Then, we find the uplink RB allocation to minimize the FL loss function.

Optimal Transmit Power

The optimal transmit power of each user i can be determined by the following proposition.

Proposition 3.1 *Given the uplink RB allocation vector r_i of each user i, the optimal transmit power of each user i, P_i^* is given by*

$$P_i^*(r_i) = \min\{P_{\max}, P_{i,\gamma_E}\}, \tag{3.61}$$

where P_{i,γ_E} satisfies the equality $\varsigma\omega_i\vartheta^2 Z(w_i) + \dfrac{P_{i,\gamma_E} Z(w_i)}{c_i^U(r_i, P_{i,\gamma_E})} = \gamma_E$.

Proof See in [29].

From Proposition 3.1, we see that the optimal transmit power depends on the size of the local FL model $Z(w_i)$ and the interference in each RB. In particular, as the size of the local FL model increases, each user must spend more energy for training FL model and, hence, the energy that can be used for data transmission decreases. In consequence, the training loss increases. Hereinafter, for simplicity, P_i^* is short for $P_i^*(r_i)$.

Optimal Uplink Resource Block Allocation

Based on Proposition 3.1 and (3.48), the optimization problem in (3.60) can be simplified as follows

$$\min_{\boldsymbol{R}} \sum_{i=1}^{U} K_i \left(1 - \sum_{n=1}^{R} r_{i,n} + \sum_{n=1}^{R} r_{i,n} q_{i,n} \right), \tag{3.62}$$

s. t. (3.52a), (3.52b), and (3.52e),

$$l_i^U(r_i, P_i^*) + l_i^D \le \gamma_T, \quad \forall i \in \mathcal{U}, \tag{3.62a}$$

$$e_i(r_i, P_i^*) \le \gamma_E, \quad \forall i \in \mathcal{U}. \tag{3.62b}$$

Obviously, the objective function (3.62) is linear, the constraints are non-linear, and the optimization variables are integers. Hence, problem (3.62) can be solved by using a standard Hungarian algorithm [87] . When the optimal RB allocation vector r_i^* is determined, the optimal transmit power of each device can be determined by (3.61) and the optimal user selection can be determined by $a_i^* = \sum_{n=1}^{R} r_{i,n}^*$. Algorithm 1 summarizes the entire process of optimizing the user selection vector $\boldsymbol{a}$, RB allocation matrix $\boldsymbol{R}$, and the transmit power vector $\boldsymbol{P}$ for training the FL algorithm.

3.3.5   *Numerical Results*

Next, we use several simulation results to show the performance of the proposed
FL algorithm. In simulations, we consider a circular network area having a radius
$r = 500$ m with one BS at its center servicing $U = 15$ uniformly distributed users.
The FL algorithm is simulated by using the Matlab Machine Learning Toolbox for
linear regression and handwritten digit identification. For linear regression, each
user implements a feedforward neural network (FNN) that consists of 20 neurons.
The data used to train the FL algorithm is generated randomly from [0, 1]. The input
x and the output y follow the function $y = -2x + 1 + n \times 0.4$ where n follows a
Gaussian distribution $\mathcal{N}(0, 1)$. The loss function is mean squared normalized error.
For handwritten digit identification, each user trains an FNN that consists of 50
neurons using the MNIST dataset [88]. The loss function is cross entropy loss. For
comparison purposes, we use three baselines: (a) an FL algorithm that optimizes
user selection with random resource allocation, (b) an FL algorithm that randomly
determines user selection and resource allocation, which can be seen as a standard
FL algorithm (e.g., similar to the one in [72]) that is not wireless-aware, and (c)
a wireless optimization algorithm that minimizes the sum packet error rates of all
users via optimizing user selection, transmit power while ignoring FL parameters.

In Fig. 3.12, we show how the identification accuracy changes as the number of
iterations varies. From Fig. 3.12, we see that, as the number of iterations increases,
the identification accuracy of all considered learning algorithms decreases first
and, then remains unchanged. The fact that the identification accuracy remains

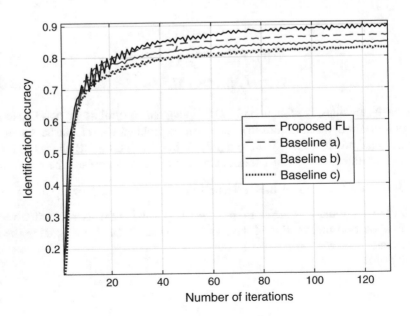

Fig. 3.12   Identification accuracy as the number of iterations varies

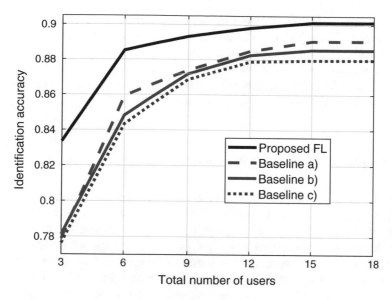

Fig. 3.13 Identification accuracy as the total number of users varies ($R = 12$)

unchanged demonstrates that the FL algorithm converges. From Fig. 3.12, we can also see that the increase speed in the value of identification accuracy is different during each iteration. This is due to the fact that the local FL models that are received by the BS may contain data errors and the BS may not be able to use them for the update of the global FL model. In consequence, at each iteration, the number of local FL models that can be used for the update of the global FL model will be different. Figure 3.12 also shows that a gap exists between the FL algorithm that optimizes the wireless factors and baselines (a), (b), and (c). This gap is caused by the packet errors.

Figure 3.13 shows how the identification accuracy changes as the total number of users varies. In this figure, an appropriate subset of users is selected to perform the FL algorithm. From Fig. 3.13, we can observe that, as the number of users increases, the identification accuracy increases. This is due to the fact that an increase in the number of users leads to more data available for the FL algorithm training and, hence, improving the accuracy of approximation of the gradient of the loss function. Figure 3.13 also shows that the FL algorithm that optimizes wireless factors improves the identification accuracy by, respectively, up to 1.2%, 1.7%, and 2.3% compared to baselines (a), (b), and (c) as the network consists of 18 users. The 1.2% improvement stems from the fact that the FL algorithm optimizes the resource allocation. The 1.7% improvement stems from the fact that the wireless aware FL algorithm joint considers learning and wireless effects and, hence, it can optimize the user selection and resource allocation to reduce the FL loss function. The 2.3% improvement stems from the fact that the proposed algorithm optimizes wireless factors while considering FL parameters such as the number of training data samples. Figure 3.13 also shows that when the number of users is less than 12, the value of the identification accuracy increases quickly. In contrast, as the number of users continues to increase, the identification accuracy increases slowly. This is because, for a higher number of users, the BS will have enough data samples to accurately approximate the gradient of the loss function.

3.4 Collaborative Federated Learning

In this section, we first provide a detailed overview of centralized learning (CL), original FL (OFL), and collaborative FL (CFL) [89], and summarize their advantages, drawbacks, and operation conditions. Then, we introduce four important performance metrics to quantify the CFL performance over IoT systems. Further, we introduce several important communication techniques to optimize the CFL performance metrics. For each communication technique, we first introduce the motivation for optimizing CFL performance and then present future research opportunities.

3.4.1 Motivation

Machine learning finds a wide range of applications in wireless networks ranging from data analytics to network monitoring and optimization [66]. However, centralized ML requires edge devices to transmit their data to a central controller for learning. In practical deployments of ML in wireless systems, such as the Internet of Things (IoT), due to privacy issues and stringent resource (e.g., bandwidth and power) constraints, edge IoT devices may not be able or willing to share their collected data with other devices or a central controller. To enable edge devices in a wireless network in training a shared ML model without data exchanges, federated learning (FL) was proposed by Google [67].

FL is a distributed ML scheme that allows IoT devices to collaboratively perform on-device training of a shared ML task while only exchanging model parameters with a central controller. Keeping the data at IoT devices not only preserves privacy but may also reduce network congestion. Due to the unique features of FL, a number of existing works (e.g., see [90, 90–93]), studied its use for wireless network optimization.

In practice, to implement FL over IoT networks, edge devices must repeatedly transmit their trained ML models to a central controller via wireless links. Due to limited wireless resources in an IoT, only a subset of devices can use FL. Meanwhile, ML models that are transmitted from IoT devices to a central controller (e.g., a base station) are subject to errors and delays caused by the wireless channel, which affects the learning performance. Therefore, it is necessary to consider the optimization of wireless networks to improve FL performance, as pointed out in [81, 82, 94–96]. This emerging *"communications for FL"* research area is the key focus of this work.

Recently, several surveys and tutorials related to FL over wireless networks appeared in [90] and [77, 97, 98]. First, the works in [77, 90, 97] investigated the use of FL for communications, rather than the impact of wireless networking on FL. Moreover, all prior works in [77, 90, 97] and [98] focused on the original FL from Google in [67] (called *original FL* hereinafter), which requires all IoT devices to transmit their ML models to a central controller. Hence, these existing surveys did not consider the implementation of FL with less or even no reliance on the central controller. Furthermore, they did not analyze how to use wireless techniques to optimize FL performance.

Next, we introduce a novel FL framework, dubbed *collaborative FL*, that combines collaborative learning with federated learning so as to enable edge devices to engage in FL without connecting to a central controller. To introduce this new framework, we first provide a detailed overview on centralized learning (CL), original FL (OFL), and collaborative FL (CFL), and summarize their advantages, drawbacks, and usage. Then, we introduce three important performance metrics to quantify the CFL performance over large-scale wireless networks. We then introduce several important communication techniques ranging from network formation, device scheduling, mobility management, and coding to optimize the CFL performance metrics. For each communication technique, we introduce the motivation for optimizing the CFL performance and then present an illustrative example and future research opportunities.

3.4.2 Preliminaries and Overview

Next, we introduce the basic architectures and differences between CL, OFL, and CFL.

As shown in Fig. 3.14a, CL needs only one ML model located at a base station (BS) or IoT cloud which works as a central controller. All devices must connect and send their data to the BS for training this ML model. Then, the BS will transmit the trained ML model to all devices. Hence, CL only requires the BS to communicate with all devices once so as to collect all devices' datasets.

The key advantage of CL is that it enables the BS or cloud to directly find a globally optimal ML model that minimizes the learning loss function value. Since the entire training process is completed by the BS, the ML training will not be affected by wireless network performance. However, imperfect wireless transmissions may introduce errors to the data used for training. Moreover, CL requires devices to transmit their collected data to the BS which leads to information leakage. In addition, significant overhead and resources are needed at the network and device levels to execute CL.

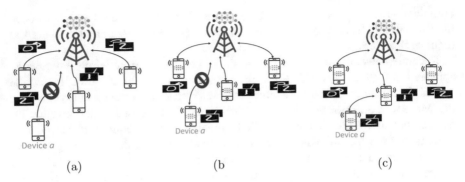

(a)   (b   (c)

Fig. 3.14 Architectures of centralized learning, original FL, and collaborative FL. (**a**) Architecture of CL. (**b**) Architecture of OFL. (**c**) Architecture of CFL

Original Federated Learning

To maintain privacy, Google's OFL framework allows each edge device to cooperatively train a shared ML model without data transmission. In OFL, both devices and the BS own an ML model with the same architecture, as shown in Fig. 3.14b. OFL is trained by an iterative learning process. First, all devices use their local data to train their local ML models and transmit their trained models to the BS. Then, the BS aggregates the received ML models, generated a new aggregate ML model, and transmits it back to all devices. Hereinafter, the ML model that is trained by an edge device is called *local FL model* while the ML model generated by the BS is called *global FL model*. At convergence, the global FL model will be equal to all local FL models, which means that devices find a shared FL model and the local FL model at convergence can be used to analyze all devices' datasets.

The key advantage of OFL is that it preserves data privacy and can be implemented over devices with less overhead than centralized ML. However, OFL still requires all devices to transmit their local FL model parameters to a BS. Hence, imperfect and dynamic wireless transmission will significantly impact the convergence time and the performance of OFL.

Collaborative Federated Learning

OFL requires all devices to send their local models to a BS, however, in practical IoT systems, devices may not be able to connect to the BS due to energy limitations or to a potentially high transmission delay. To overcome this challenge and facilitate the use of OFL in real-world IoT systems, we propose the concept of CFL using which devices can engage in FL without connecting to a BS or a cloud.

In CFL, devices that cannot connect to the BS directly can associate with neighboring users. For example, as shown in Fig. 3.14b, for OFL, device *a* cannot connect to the BS and perform FL due to a potentially high transmission delay.

However, in CFL, as shown in Fig. 3.14c, device a can connect to its closest device for performing FL. CFL is also trained iteratively. First, each device transmits its trained local FL model to its connected devices or the BS. Then, the BS generates the global FL model and transmits it to the associated devices. Finally, each device updates its local FL model based on the local FL models received from other devices or the BS. In OFL, each device must train its local FL model using gradient descent (GD) methods while the BS aggregates the local FL models. However, in CFL, each device must both aggregate the local FL models received from other devices and train its local FL model.

To show the difference between CFL and OFL, we implemented a preliminary simulation for a network having one BS and six devices, as shown in Fig. 3.15a. The local FL model of each device consists of a shallow feedforward neural network with 50 neurons. The MNIST dataset [88] is used for training the local FL models at each device and each device has 500 data samples. OFL is used for comparison. The maximum time used for FL model parameter transmission is set to be 0.23 s.

Figure 3.15b shows how the identification accuracy changes over time. Figure 3.15b demonstrates that CFL outperforms OFL. This is because, for OFL, only four devices can participate in FL and the other two devices have a delay larger than 0.23 s. Since CFL allows devices to connect to other devices and the transmission

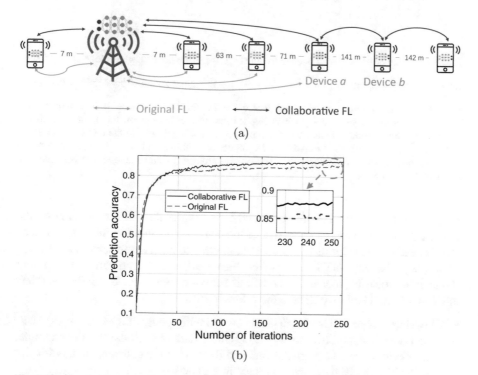

Fig. 3.15 Simulation system and result to show the performance of CFL and OFL. In this figure, a red digit is the distance between two adjacent devices. (**a**) Simulation system. (**b**) Simulation result

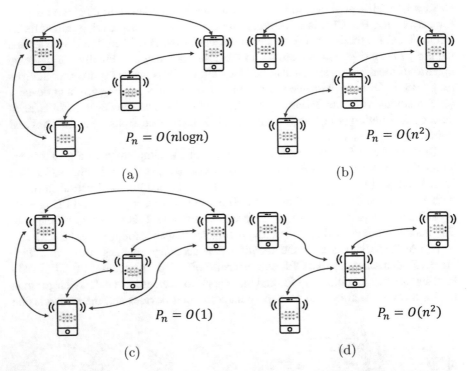

Fig. 3.16 Number of iterations needed to converge for different CFL algorithms with different topologies. In this figure, $O\left(\frac{\max\left((g^0-w^*)^4, L^4 P_n^2\right)}{\varepsilon^2}\right)$ is the upper bound of the number of iterations that a CFL algorithm needs to converge, where n is the number of devices that perform the FL algorithm, ε is the target accuracy which implies the difference between the optimal FL model and the FL model at convergence, L is the upper bound of the gradient of the loss function, $g^0 = \frac{1}{n}\sum_{i=1}^{n} w_i^0$ with w_i^0 being the initial local FL model of device i, and w^* is the optimal local FL model at convergence. (**a**) Grid topology. (**b**) Path topology. (**c**) Complete topology. (**d**) Complete topology Star topology

delay between any two neighboring devices is smaller than 0.23 s, six devices can participate. In fact, CFL can also reduce the energy consumption for device b since it only needs to transmit its ML model parameters to device a instead of the BS.

The key advantage of CFL is that it enables the devices to perform the FL without transmitting local FL models to the BS, as shown in Fig. 3.16. Given the overview of CL, OFL, and CFL, we remark the following:

- Choosing between CL or FL depends on: (a) willingness of data sharing, (b) ML model data size, and (c) size of the collected data of each device. For example, when devices agree to share the data and the size of the collected data is smaller than the ML model data size, CL is recommended.

- Choosing between OFL or CFL depends on: (a) whether the BS performs FL and (b) the connection and transmission delay between devices and the BS. For example, if all IoT devices need to implement FL without the BS, then CFL is more suitable.
- OFL can be considered as a special case of CFL. In a network, if each device connects to all other devices, CFL is equivalent OFL.

3.4.3 Communication Techniques for Collaborative Federated Learning

We now overview key techniques that can be used to improve the performance of CFL over wireless networks.

Network Formation

The first fundamental step towards deploying CFL is to analyze the process of network formation using which devices can connect to one another to engage in a CFL task. In CFL, devices can form different network topologies. For example, IoT devices can form a grid topology for CFL, as shown in Fig. 3.16a. Naturally, the training complexity and the FL convergence time directly depend on the formed topology. Hence, for any given network scenario, it will be interesting to investigate the optimal CFL network topology.

Figure 3.16 shows the upper bound of the number of iterations for CFL convergence when assuming that the upper bound is derived based on the assumption that each device updates its local FL model using the Lazy Metropolis method. Figure 3.16 shows that, when the number of links of each device increases, the number of iterations decreases because having more links increases the frequency of local FL model sharing.

Clearly, CFL yields interesting network formation research questions as follows:

- **Optimal CFL network formation**: The optimal CFL network topology depends on the CFL performance metrics being optimized. Therefore, a fundamental CFL question is that of network formation: How can the devices interact to form an optimal network topology that maximizes the various CFL performance metrics and tradeoffs? To find the optimal CFL network topology, the first step is to define a proper utility function that jointly considers multiple dependent CFL performance metrics and network topology. Given the defined utility function, one must develop network formation algorithms to optimize the utility function. Both centralzied and distributed solutions can be developed. Centralized solutions such as searching based algorithms may be able to find the globally optimal network topology. However, the implementation of centralized solutions requires all devices' information such as locations or wireless channel conditions, which is impractical for a large-scale and dynamic IoT system. For distributed solutions, one can adopt a game-theoretic approach, particularly using network formation games [56]. In network formation games, each device is seen as an individual agent whose goal is to form a graph with neighboring devices so as to optimize the CFL performance metrics. The CFL performance (e.g., utility) depends on the entire graph and decision of all agents which makes the use of game theory suitable. One unique feature of the CFL network formation game is that it could be dynamic and requires far-sighted decision making. That is an angle that has only been studied in limited prior works as discussed in [56].

- **Network formation with asynchronous training**: Under asynchronous FL training, IoT devices will update and transmit their local FL models at different time slots. Due to limited computing and wireless resources, each device may not want to update its local FL model until it receives all local FL models of its associated devices. Using asynchronous training can increase local FL model update frequency and the data rate of each device which reduces the convergence time. In asynchronous training, the number of devices that need to transmit the local FL models is time-varying. Hence, the network topology must be adapted to the changes in the number of devices that must transmit local FL models. Here, one must determine the frequency with which the network topology must be updated according to the number of participating devices. Note that each network topology update will change the wireless resource allocation and device association schemes so as to improve CFL performance metrics such as convergence time. However, network topology updates will also introduce communication overhead such as network state information sharing.

- **Network formation with partial network information**: In actual IoT, each device may not completely know the network architecture, device locations, and network composition. Due to this limited information, the number of devices that each device can connect to is limited and hence devices may not be able to form a network topology that satisfies the CFL usage conditions. Therefore, there is a need to investigate a globally optimal network formation for IoT devices with partial information. Since most existing complexity results related to network formation (e.g., see [99]) assume that each device has complete information, they cannot be used for devices with partial network information. Meanwhile, due to partial network information, devices may form several unconnected small device groups. Hence, a multi-layer network formation must be designed. For example, in the first layer, devices will exchange their local FL model parameters in their own groups while the local FL model parameters are exchanged over multiple groups in the second layer. The designed scheme must balance the communication overheads and training complexity among multiple layers.

Device Scheduling

Due to energy constraints and wireless resource limitations, the number of devices that can engage in CFL is limited. Hence, an IoT device may update its local FL model using the local FL models of a subset of devices thus decreasing the CFL convergence time. Therefore, it is necessary to find an optimal device scheduling policy that can determine the frequency and which iterations that each device engages in CFL so as to optimize the CFL performance metrics.

Device scheduling plays an important role in training CFL and it also faces several interesting research problems:

- **Data importance-aware device scheduling**: In CFL, the contribution of each device's dataset on the update of a local FL model can be seen as the data importance of that device's dataset. The data importance of each device depends on the number of training data samples and the data distribution. For instance, if a device has a large number of training data samples, its local FL model will be allocated a large weight within the local FL model update. Since only a subset of devices can perform FL at each iteration, it is necessary to design data importance-aware device scheduling policies for improving convergence speed. In particular, one must first build a data importance model that jointly considers the number of training data samples, data distribution, and data uniqueness. Meanwhile, in CFL, devices cannot share data and, hence, each device may not be able to directly know the data importance of other devices. Therefore, there is a need to find a method to learn the data importance of other devices from their transmitted local FL model parameters. In addition, one must determine the frequency of local FL model update for devices with different data importance. Note that increasing the update frequency of the devices with high data importance can improve convergence speed but it also increases the loss function value.
- **Device scheduling for multiple FL tasks**: In a wireless network, a device may perform multiple FL algorithms simultaneously. Therefore, it will be interesting to design a device scheduling policy that enables devices to efficiently train multiple FL models and transmit the trained FL models to other devices simultaneously. Since each FL task has its specific convergence time requirement and target loss function value, the developed device scheduling policy must determine which FL model must be trained first and which FL model must be transmitted first so as to satisfy the requirements of each FL task. Moreover, since the convergence time of each FL task is different, the designed scheduling policy must be adapted to the changes in the number of incomplete FL tasks.
- **Device scheduling and network formation for mobile devices**: In an IoT system, several devices, such as cars and drones, are mobile. The connections among different devices and the wireless network performance will change depending on the mobility of the devices thus affecting the CFL performance. Meanwhile, device mobility will increase the frequency of devices changing their connections thus slowing down the CFL training process. Therefore, it is

necessary to study device scheduling and network formation for mobile devices. In OFL, devices will transmit their local FL models to a static BS. However, in CFL, mobile devices must transmit their local FL models to other mobile devices. Hence, the devices' locations and connections are correlated in space (i.e., between two connected devices) and time (i.e., between time slots). For example, for two devices moving in parallel, although the location will be changing, the distance between the two devices remains constant. As a result, the change of their locations will not increase the local FL model transmission delay. Therefore, one must first build a model to capture the effect of spatio-temporal correlation of device locations and connections on the FL performance metrics. Then, it must investigate how to use spatio-temporal correlation to optimize device scheduling and network topology policies and the frequency of changing these policies.

Coding

During the CFL training process, source coding, channel coding, and gradient coding can be used to improve the FL performance. Source coding is used to compress the high-dimensional FL model parameters so that they can be represented by a small number of bits hence reducing the FL parameter transmission delay. Channel coding is used to protect the transmitted FL model parameters against the wireless noise and interference thus improving packet error rates and CFL reliability. Gradient coding is used to encode the gradient descent parameters of machine learning algorithms so as to improve the ML performance.

Obviously, source, channel, and gradient coding can significantly improve CFL performance. However, a number of research questions still exists:

- **Heterogeneous source coding design**: In an IoT system, the wireless transmission link characteristics of each device will be different (e.g., different data rates). To efficiently use wireless resources for FL model transmission, each device may encode its local FL model using different number of bits or different coding techniques. This type of coding schemes is called heterogeneous source coding. For example, some devices can use 15 bits to represent their local FL models while another can use 7 bits to represent its local FL model. Heterogeneous source coding can significantly reduce the coding energy consumption and decrease the loss function value. However, in CFL, a device must transmit its local FL model to multiple devices. Therefore, one must determine the number of local FL models that each device must encode and the number of bits used to encode the corresponding local FL models. For example, if a given device must transmit its local FL model to three devices, then this device can encode a local FL model and transmit it to three devices. Also, the device can encode two or three local FL models with different number of bits and then transmit them to these three devices.
- **Gradient coding for avoiding stragglers**: Due to limited wireless resources, an IoT system has devices with extremely high transmission delay or computational

delay. Such devices (called stragglers) may not be able to complete the local FL model transmission within the time duration required by the system. If a network has a large number of stragglers, the number of devices that can perform CFL will significantly decrease. Therefore, there is a need to design gradient coding schemes for addressing the problem of stragglers. However, traditional gradient coding methods require devices to share their dataset with other devices so as to remove stragglers and hence, they cannot be used for CFL since CFL does not allow devices to share their data. Hence, one must investigate a novel gradient coding scheme without data sharing.

3.5 Summary

In this chapter, we have presented a joint learning and communication framework for federated learning over wireless networks. In first part, we have formulated an optimization problem that jointly considers user selection and resource allocation for the minimization of federated learning training loss. To solve this problem, we have derived a closed-form expression for the expected convergence rate of the FL algorithm that considers the limitations of the wireless medium. Based on the derived expected convergence rate, the optimal transmit power is determined given the user selection and uplink resource block allocation. Then, the Hungarian algorithm is used to find the optimal user selection and RB allocation so as to minimize the FL loss function. Simulation results have shown that the joint federated learning and communication framework yields significant improvements in the performance compared to existing implementations of the FL algorithm that does not account for the properties of the wireless channel.

In the next part, we have formulated a Federated Learning over wireless network as an optimization problem that captures both trade-offs, such as (i) between computation and communication latencies determined by learning accuracy level, and thus (ii) between the Federated Learning time and user equipment energy consumption. By decomposing the problem into three sub-problems with convex structure, we then characterized how the computation and communication latencies of mobile devices affect to various trade-offs between the user equipment energy consumption, system learning time, and learning accuracy parameter, and also quantified the impact of user equipment heterogeneity on the system cost. In the final part, we have presented a novel collaborative federated learning framework that enables the participation of end-devices having limited communication resources for performance improvement. Collaborative federated learning has shown performance improvement in terms of learning accuracy. Therefore, one can use collaborative federated learning for various IoT applications.

Chapter 4
Incentive Mechanisms for Federated Learning

Abstract In this chapter, we discuss various components of federated learning that must be given some incentive in terms of monetary cost or other benefits. We design incentive mechanism design for federated learning using game theory and auction theory. Finally, we present extensive numerical results to show the validity of our proposed incentive mechanisms.

4.1 Introduction

Federated learning can be trained mainly in two different ways, such as (a) centralized server aggregation-based training, and (b) blockchain-based training, as shown in Figs. 4.1 and 4.2. Federated learning using centralized server aggregation involves continuous, iterative interaction between the end-devices and aggregation server. End-devices use their resources (i.e., computation resource and energy) to train their local learning models. The locally trained model updates will be sent via a wireless channel to the aggregation server for global aggregation. Similar to end-devices, the aggregation server will use its resources (i.e., computation resource and energy) to perform aggregation. To enable successful interaction among end-devices and aggregation servers for federated learning requires an attractive incentive mechanism. End-devices must be provided with benefits in response to their participation in the federated learning process. On the other hand, blockchain-based federated learning involves the computation of local models at the end-devices. The end-devices send their local learning models to their corresponding miners. The miners perform sharing and cross-verification of learning models to avoid injection of wrong models. Then, all the miners start computing their consensus algorithms (e.g., Proof-of-Work). The winning miner that solves the consensus algorithm first, broadcasts its block to all the miners in the network for updating their blocks. In blockchain-based federated learning, there is a need to provide an attractive incentive to both end-devices and miners for their jobs. Therefore, the incentive mechanism for blockchain-based federated learning will be different than the one for federated learning based on a centralized aggregation server.

© The Author(s), under exclusive license to Springer Nature Singapore Pte Ltd. 2021
C. S. Hong et al., *Federated Learning for Wireless Networks*, Wireless Networks,
https://doi.org/10.1007/978-981-16-4963-9_4

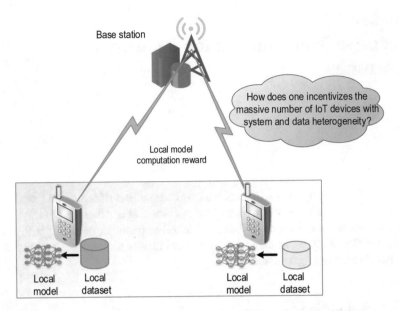

Fig. 4.1 Overview of rewards in centralized aggregation-based federated learning

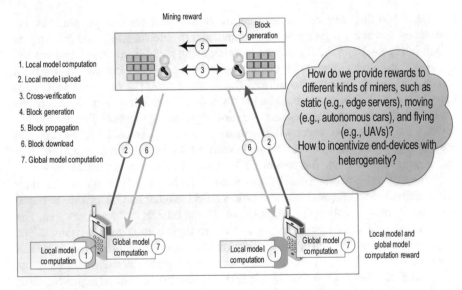

Fig. 4.2 Overview of rewards in blockchain-based federated learning

Generally, we can categorize incentives into two main types: monetary and non-monetary [100]. Monetary incentives are based on providing end-devices with payments as per their participation, whereas non-monetary incentives generally involve providing end-devices with benefits other than payments. Non-monetary incentives in the case of federated learning can be the well-trained global federated learning model for a large number of end-devices. Unless stated otherwise, the keyword incentive in this chapter refers to monetary incentive. Next, we present incentive mechanisms based on game theory and auction theory for federated learning over wireless networks.

4.2 Game Theory-Enabled Incentive Mechanism

Game theory has proven to be one of the successful tools in enabling/optimizing various functions/design aspects in wireless networks, such as wireless resource allocation, computational offloading in edge computing, URLLC/eMBB coexistence, and incentive mechanism design, among others [101–105]. Generally, games can be divided into (a) cooperative and (b) non-cooperative games. Cooperative games are based on achieving the equilibrium state for optimizing the overall benefit via joint decision-making by the various players. On the other hand, the players in non-cooperative games choose their strategies selfishly without coordination with other players (Fig. 4.3). A summary of cooperative and non-cooperative games used for incentive mechanism design in wireless networks is given in Table 4.1.

In federated learning, local computations at the devices and their communication with the centralized coordinating server are interleaved in a complex manner to build a global learning model. Therefore, a communication-efficient federated learning framework [18, 58] requires solving several challenges. Furthermore, because of limited data per device to train a high-quality learning model, the difficulty is to incentivize a large number of mobile users to ensure cooperation. This important aspect in federated learning has been overlooked so far, where the question is *how can we motivate a number of participating clients, collectively providing a large number of data samples to enable federated learning without sharing their private data?* Note that both participating clients and the server can benefit from training a global model. However, to fully reap the benefits

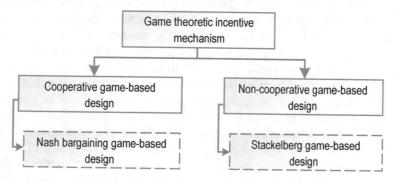

Fig. 4.3 Classification of game theoretic incentive mechanisms

Table 4.1 Overview of game theoretic incentive mechanisms [106]

Reference	Category	Game	Primary focus
Ho et al. [105]	Non-cooperative game	Stackelberg game	Macrocell base station traffic is admitted by the small cell base stations for monetary benefits
Liu et al. [107]	Non-cooperative	Stackelberg game	Incentives are given to femto cells for sharing their resource to macrocell base stations
Poularakis et al. [108]	Non-cooperative game	Stackelberg game	Cache-enabled access points are incentivized for caching by macrocell base stations
Gao et al. [109]	Cooperative game	Nash bargaining game	A cooperative game is proposed to carryout the transactions between the single mobile virtual network operator and multiple access points.
Yu et al. [110]	Cooperative game	Nash bargaining game	A bargaining framework in which mobile network operator bargains with venue owners sequentially for determining the deployment locations of Wi-Fi and how much to pay.

of high-quality updates, the multi-access edge computing (MEC) server has to incentivize clients for participation. In particular, under heterogeneous scenarios, such as an adaptive and cognitive-communication network, the client's participation in federated learning can spur collaboration and provide benefits for operators to accelerate and deliver network-wide services [111]. Similarly, clients, in general, are not concerned with the reliability and scalability issues of federated learning [112]. Therefore, to incentivize users to participate in the collaborative training, we require a marketplace. For this purpose, we present a value-based compensation mechanism to the participating clients, such as a bounty (e.g., data discount package), as per their level of participation in the crowdsourcing framework. This is reflected in terms of local accuracy level, i.e., *quality of solution to the local subproblem*, in which the framework will protect the model from imperfect updates by restricting the clients trying to compromise the model (for instance, with skewed data because of its i.i.d nature or data poisoning) [113]. Moreover, we cast the global loss minimization problem as a primal-dual optimization problem, instead of adopting a traditional gradient descent learning algorithm in the federated learning setting (e.g., FedAvg [18]). This enables (a) proper assessment of the quality of the local solution to improve personalization and fairness amongst the participating clients while training a global model, (b) effective decoupling of the local solvers, thereby balancing communication and computation in the distributed setting.

The goal of this section is two-fold: First, we formalize an incentive mechanism to develop a participatory framework for mobile clients to perform federated learning for improving the global model. Second, we address the challenge of maintaining communication efficiency while exchanging the model parameters with a number of participating clients during aggregation. Specifically, communication efficiency in this scenario accounts for communications per iteration with an arbitrary algorithm to maintain an acceptable accuracy level for the global model. In this work, we design and analyze a novel crowdsourcing framework to realize the federated learning vision. Specifically, our contributions are summarized as follows:

- **A crowdsourcing framework to enable communication -efficient federated learning**. We design a crowdsourcing framework, in which federated learning participating clients iteratively solve the local learning subproblems for an accuracy level subject to an offered incentive. We then establish a communication-efficient cost model for the participating clients. We then formulate an incentive mechanism to induce the necessary interaction between the MEC server and the participating clients for the federated learning in Sect. 4.2.2.
- **Solution approach using Stackelberg game**. With the offered incentive, the participating clients independently choose their strategies to solve the local subproblem for a certain accuracy level in order to minimize their participation costs. Correspondingly, the MEC server builds a high-quality centralized model characterized by its utility function, with the data distributed over the participating clients by offering the reward. We exploit these tightly coupled motives of the participating clients and the MEC server as a two-stage Stackelberg game. The equivalent optimization problem is characterized as mixed-Boolean programming which requires an exponential complexity effort for finding the solution. We analyze the game's equilibria and propose a linear complexity algorithm to obtain the optimal solution.
- **Participant's response analysis and case study**. We next analyze the response behavior of the participating clients via the solutions of the Stackelberg game and establish the efficacy of our proposed framework via case studies. We show that the linear-complexity solution approach attains the same performance as the mixed-Boolean programming problem. Furthermore, we show that our mechanism design can achieve the optimal solution while outperforming a heuristic approach for attaining the maximal utility with up to 22% of gain in the offered reward.
- **Admission control strategy**. Finally, we show that it is significant to have certain participating clients to guarantee the communication efficiency for an accuracy level in federated learning. We formulate a probabilistic model for threshold accuracy estimation and find the corresponding number of participants required to build a high-quality learning model. We analyze the impact of the number of participants in federated learning while determining the threshold accuracy level with closed-form solutions. Finally, with numerical results, we demonstrate the structure of the admission control model for different configurations.

4.2.1   System Model

Figure 4.4 illustrates our proposed system model for the crowdsourcing framework to enable federated learning. The model consists of a number of mobile clients associated with a base station having a central coordinating server (MEC server), acting as a central entity. The server facilitates the computation of the parameters aggregation, and feedback the global model updates in each global iteration. We consider a set of participating clients $\mathcal{K} = \{1, 2, \ldots, K\}$ in the crowdsourcing framework. The crowdsourcer (platform) can interact with mobile clients via an application interface and aims at leveraging federated learning to build a global ML model. As an example, consider a case where the crowdsourcer (referred to as MEC server hereafter, to avoid any confusion) wants to build a ML model. Instead of just relying on available local data to train the global model at the MEC server, the global model is constructed utilizing the local training data available across several distributed mobile clients. Here, the global model parameter is first shared by the MEC server to train the local models in each participating client. The local model's parameters minimizing local loss functions are then sent back as feedback and are aggregated to update the global model parameter. The process continues iteratively, until convergence.

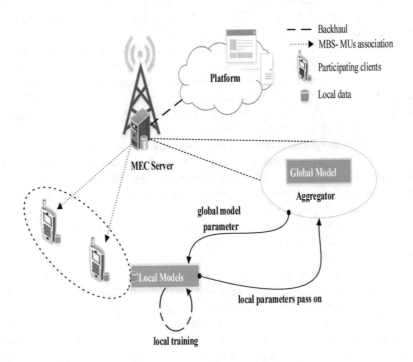

Fig. 4.4 Crowdsourcing framework for decentralized machine learning

---

Algorithm 4 Federated learning framework

1: **Input:** Initialize dual variable $\alpha^0 \in \mathbb{R}^D$, D_k, $\forall k \in K$.
2: **for** Each aggregation round **do**
3:     **for** $k \in \mathcal{K}$ **do**
4:         Solve local subproblems (4.5) in parallel.
5:         Update local variables as in (4.7).
6:     **end for**
7:     Aggregate to update global parameter as in (4.8).
8: **end for**

Federated Learning Background

For federated learning, we consider unevenly partitioned training data over a large number of participating clients to train the local models under any arbitrary learning algorithm. Each client k stores its local dataset $\mathcal{D}_k$ of size D_k respectively. Then, we define the training data size $D = \sum_{k=1}^{K} D_k$. In a typical supervised learning setting, $\mathcal{D}_k$ defines the collection of data samples given as a set of input-output pairs $\{x_i, y_i\}_{i=1}^{D_k}$, where $x_i \in \mathbb{R}^d$ is an input sample vector with d features, and $y_i \in \mathbb{R}$ is the labeled output value for the sample x_i. The learning problem, for an input sample vector x_i (e.g., the pixels of an image) is to find the *model parameter vector* $w \in \mathbb{R}^d$ that characterizes the output y_i (e.g., the labeled output of the image, such as the corresponding product names in a store) with the loss function $f_i(w)$. Some examples of loss functions include $f_i(w) = \frac{1}{2}(x_i^T w - y_i)^2$, $y_i \in \mathbb{R}$ for a linear regression problem and $f_i(w) = \max\{0, 1 - y_i x_i^T w\}$, $y_i \in \{-1, 1\}$ for support vector machines. The term $x_i^T w$ is often called a *linear mapping function*. Therefore, the loss function based on the local data of client k, termed local subproblem is formulated as

$$J_k(w) = \frac{1}{D_k} \sum_{i=1}^{D_k} f_i(w) + \lambda g(w), \tag{4.1}$$

where $w \in \mathbb{R}^d$ is the local model parameter, and $g(\cdot)$ is a regularizer function, commonly expressed as $g(\cdot) = \frac{1}{2}\|\cdot\|^2$; $\forall \lambda \in [0, 1]$. This characterizes the *local model* in the federated learning setting.

Global Problem At the MEC server, the global problem can be represented as the finite-sum objective of the form

$$\min_{w \in \mathbb{R}^d} J(w) \quad \text{where} \quad J(w) \equiv \frac{\sum_{k=1}^{K} D_k J_k(w)}{D}. \tag{4.2}$$

Problems of such structure as in (4.2) where we aim to minimize an average of K local objectives are well-known as *distributed consensus problems* [114].

Solution Framework under Federated Learning We recast the regularized global problem in (4.2) as

$$\min_{w \in \mathbb{R}^d} J(w) := \frac{1}{D} \sum_{i=1}^{D} f_i(w) + \lambda g(w), \tag{4.3}$$

and decompose it as a dual optimization problem[1] in a distributed scenario [115] amongst K participating clients. For this, at first, we define $X \in \mathbb{R}^{d \times D_k}$ as a matrix with columns having data points for $i \in \mathcal{D}_k, \forall k$. Then, the corresponding dual optimization problem of (4.3) for a convex loss function f is

$$\max_{\alpha \in \mathbb{R}^D} \mathcal{G}(\alpha) := \frac{1}{D} \sum_{i=1}^{D} -f_i^*(-\alpha_i) - \lambda g^*(\phi(\alpha)), \tag{4.4}$$

where $\alpha \in \mathbb{R}^D$ is the dual variable mapping to the primal candidate vector, f_i^* and g^* are the convex conjugates of f_i and g respectively [116]; $\phi(\alpha) = \frac{1}{\lambda D} X \alpha$. With the optimal value of dual variable α^* in (4.4), we have $w(\alpha^*) = \nabla g^*(\phi(\alpha^*))$ as the optimal solution of (4.3) [115]. For the ease of representation, we will use $\phi \in \mathbb{R}^d$ for $\phi(\alpha)$ hereafter. We consider that g is a strongly convex function, i.e., $g^*(\cdot)$ is continuous differentiable. Then, the solution is obtained following an iterative approach to attain a global accuracy $0 \le \epsilon \le 1$ (i.e., $\mathbb{E}\left[\mathcal{G}(\alpha) - \mathcal{G}(\alpha^*)\right] < \epsilon$).

Under the distributed setting, we further define data partitioning notations for clients $k \in \mathcal{K}$ to represent the working principle of the framework. Let us define a weight vector $\varrho_{[k]} \in \mathbb{R}^D$ at the local subproblem k with its elements zero for the unavailable data points. Following the assumption of having f_i as $(1/\gamma)$-smooth and 1-strongly convex of g to ensure convergence, its consequences is the approximate solution to the local problem k defined by the dual variables $\alpha_{[k]}, \varrho_{[k]}$, characterized as

$$\max_{\varrho_{[k]} \in \mathbb{R}^D} \mathcal{G}_k(\varrho_{[k]}; \phi, \alpha_{[k]}), \tag{4.5}$$

where $\mathcal{G}_k(\varrho_{[k]}; \phi, \alpha_{[k]}) = -\frac{1}{K} - \langle \nabla(\lambda g^*(\phi(\alpha))), \varrho_{[k]} \rangle - \frac{\lambda}{2} \| \frac{1}{\lambda D} X_{[k]} \varrho_{[k]} \|^2$ is defined with a matrix $X_{[k]}$ columns having data points for $i \in \mathcal{D}_k$, and zero padded otherwise. Each participating client $k \in \mathcal{K}$ iterates over its computational resources using any arbitrary solver to solve its local problem (4.5) with a local relative θ_k accuracy that characterizes the quality of the local solution, and produces a random output $\varrho_{[k]}$ satisfying

$$\mathbb{E}\left[\mathcal{G}_k(\varrho_{[k]}^*) - \mathcal{G}_k(\varrho_{[k]})\right] \le \theta_k \left[\mathcal{G}_k(\varrho_{[k]}^*) - \mathcal{G}_k(0)\right]. \tag{4.6}$$

---

[1] The duality gap provides a certificate to the quality of local solutions and facilitates distributed training.

Note that, with local (relative) accuracy $\theta_k \in [0, 1]$, the value of $\theta_k = 1$ suggests that no improvement was made by the local solvers during successive local iterations. Then, the local dual variable is updated as follows:

$$\alpha_{[k]}^{t+1} := \alpha_{[k]}^t + \varrho_{[k]}^t, \forall k \in \mathcal{K}. \tag{4.7}$$

Correspondingly, each participating client will broadcast the local parameter defined as $\Delta\phi_{[k]}^t := \frac{1}{\lambda D}X_{[k]}\varrho_{[k]}^t$, during each round of communication to the MEC server. The MEC server aggregates the local parameter (averaging) with the following rule:

$$\phi^{t+1} := \phi^t + \frac{1}{K}\sum_{k=1}^{K} \Delta\phi_{[k]}^t, \tag{4.8}$$

and distributes the global change in ϕ to the participating clients, which is used to solve (4.5) in the next round of local iterations. This way we observe the decoupling of global model parameter from the need of local clients' data[2] for training a global model.

Algorithm 4 briefly summarizes the federated learning framework as an iterative process to solve the global problem characterized in (4.3) for a global accuracy level. The iterative process (S2)–(S8) of Algorithm 4 terminates when the global accuracy ϵ is reached. A participating client k strategically[3] iterates over its local training data $\mathcal{D}_k$ to solve the local subproblem (4.5) up to an accuracy θ_k. In each communication round with the MEC server, the participating clients *synchronously* pass on their parameters $\Delta\phi_{[k]}$ using a shared wireless channel. The MEC server then aggregates the local model parameters ϕ as in (4.8), and broadcasts the global parameters required for the participating clients to solve their local subproblems for the next communication round. Within the framework, consider that each participating client uses any arbitrary optimization algorithm (such as *Stochastic Gradient Descent (SGD), Stochastic Average Gradient (SAG), Stochastic Variance Reduced Gradient (SVRG)*) to attain a relative θ accuracy per local subproblem. Then, for strongly convex objectives, the number of iterations is dependent on local relative θ accuracy of the local subproblem and the global model's accuracy ϵ as [58]:

$$I^g(\epsilon, \theta) = \frac{\zeta \cdot \log(\frac{1}{\epsilon})}{1 - \theta}, \tag{4.9}$$

where the local relative accuracy measures the quality of the local solution as defined in the earlier paragraphs. Further, in this formulation, we have replaced the term $\mathcal{O}(\log(\frac{1}{\epsilon}))$ in the numerator with $\zeta \cdot \log(\frac{1}{\epsilon})$, for a constant $\zeta > 0$. For fixed iterations I^g at the MEC server to solve the global problem, we observe in (4.9) that

---

[2] Note that we consider the availability of quality of data with each participating client for solving a corresponding local subproblem. Further related demonstration on dependency of the normalized data size and accuracy can be found in [117].

[3] Fewer iterations might not be sufficient to have an optimal local solution [111].

a very high local accuracy (small θ) can significantly improve the global accuracy ϵ. However, each client k has to spend excessive resources in terms of local iterations, I_k^1 to attain a small θ_k accuracy as

$$I_k^1(\theta_k) = \gamma_k \log\left(\frac{1}{\theta_k}\right),$$

(4.10)

where $\gamma_k > 0$ is a parameter choice of client k that depends on the data size and condition number of the local subproblem [58]. Therefore, to address this trade-off, MEC server can setup an economic interaction environment (a crowdsourcing framework) to motivate the participating clients for improving the local relative θ_k accuracy. Correspondingly, with the increased reward, the participating clients are motivated to attain better local θ_k accuracy, which as observed in (4.9) can improve the global ϵ accuracy for a fixed number of iterations I^g of the MEC server to solve the global problem. In this scenario, to capture the statistical and system-level heterogeneity, the corresponding performance bound in (9) for heterogeneous responses θ_k can be modified considering the worst-case response of the participating client as

$$I^g(\epsilon, \theta_k) = \frac{\zeta \cdot \log(\frac{1}{\epsilon})}{1 - \max_k \theta_k}, \forall k \in \mathcal{K}.$$

(4.11)

Figure 4.5 describes an interaction environment incorporating a crowdsourcing framework and federated learning setting. In the following section, we will further discuss in detail the proposed incentive mechanism and present the interaction between the MEC server and participating clients as a two-stage Stackelberg game.

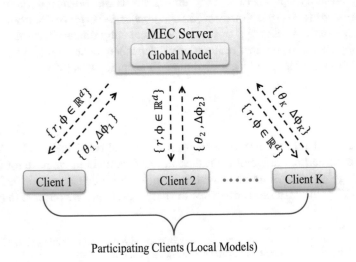

Fig. 4.5 Interaction environment of federated learning setting under crowdsourcing framework

Cost Model

Training on local data for a defined accuracy level incurs a cost for the participating clients. We discuss its significance with two typical costs: the computing cost and the communication cost.

Computing Cost This cost is related to the number of iterations performed by client k on its local data to train the local model for attaining a relative accuracy of θ_k in a single round of communication. With (4.10), we define the computing cost for client k when it performs computation on its local data $\mathcal{D}_k$.

Communication Cost This cost is incurred when client k interacts with MEC server for parameter updates to maintain θ_k accuracy. During a round of communication with the MEC server, let e_k be the size (in bits) of local parameters $\Delta\phi_{[k]}$, $k \in \mathcal{K}$ in a floating-point representation produced by the participating client k after processing a mini-batch [118]. While e_k is the same for all the participating clients under a specified learning setting of the global problem, each participating client k can invest resources to attain specific θ_k as defined in (4.10). Although the best choice would be to choose θ_k such that the local solution time is comparable with the time expense in a single communication round, larger θ_k will induce more rounds of interaction between clients until global convergence, as formalized in (4.9).

With the inverse relation of global iteration upon local relative accuracy in (4.9), we can characterize the total communication expenditure as

$$T(\theta_k) = \frac{T_k}{(1 - \theta_k)}, \tag{4.12}$$

where T_k as the time required for the client k to communicate with MEC server in each round of model's parameter exchanges. Here, we normalize $\zeta > 0$ in (4.9) to 1 as the constant can be absorbed into T_k for each round of model's parameter exchanges when we characterize the communication expenditure in (4.12). Using first-order Taylor's approximation,[4] we can approximate the total communication cost as $T(\theta_k) = T_k \cdot (1 + \theta_k)$. We assume that clients are allocated orthogonal sub-channels so that there is no interference between them.[5] Therefore, the instantaneous data rate for client k can be expressed as

$$R_k = B \log_2 \left(1 + \frac{p_k |G_k|^2}{\mathcal{N}_k}\right), \forall k \in \mathcal{K}, \tag{4.13}$$

---

[4] First-order Taylor's approximation for $f(\theta) = \frac{1}{1-\theta}$ is $f(\theta)\,|_{\theta=a} = f(a) + f'(a)(\theta - a)$. For small θ, the approximation results $f(\theta)\,|_{\theta=0} = 1 + \theta$.

[5] Note that the scenario of possible delay introduced with interference on poor wireless uplink channel can affect the local model update time. This can be mitigated by adjusting maximum waiting time as in [112] at MEC.

where B is the total bandwidth allocated to the client k, p_k is the transmission power of the client k, $|G_k|^2$ is the channel gain between participating client k and the base station, and $\mathcal{N}_k$ is the Gaussian noise power at client k. Then for client k, using (4.13), we can characterize T_k for each round of communication with the MEC server to upload the required updates as

$$T_k = \frac{e_k}{B \log_2 \left(1 + \frac{p_k |G_k|^2}{\mathcal{N}_k} \right)}, \forall k \in \mathcal{K}. \tag{4.14}$$

(4.14) provides the dependency of T_k on wireless conditions and network connectivity.

Assimilating the rationale behind our earlier discussions, for a participating client with evaluated T_k, the increase in value of θ_k (poor local accuracy) will contribute to a larger communication expenditure. This is because the participating client has to interact more frequently with the MEC server (increased number of global iterations) to update its local model parameter for attaining relative θ_k accuracy. Further, the authors in [119] have provided the convergence analysis to justify this relationship and the communication cost model, though with a different technique.

Therefore, the participating client k's cost for the relative accuracy level θ_k on the local subproblem is

$$C_k(\theta_k) = (1 + \theta_k) \cdot \left(v_k \cdot T_k + (1 - v_k) \cdot \gamma_k \log \left(\frac{1}{\theta_k} \right) \right), \tag{4.15}$$

where $0 \leq v_k \leq 1$ is the normalized monetary weight for communication and computing costs (i.e., \$/ rounds of iteration). A **smaller value of relative accuracy** θ_k **indicates a high local accuracy**. Thus, there exists a trade-off between the communication and the computing cost (4.15). A participating client can adjust its preference on each of these costs with the weight metric v_k. The higher value of v_k emphasizes the larger rounds of interaction with the MEC server to adjust its local model parameters for the relative θ_k accuracy. On the other hand, the higher value of $(1-v_k)$ reflects the increased number of iterations at the local subproblem to achieve the relative θ_k accuracy. This will also significantly reduce the overall contribution of communication expenditure in the total cost formulation for the client. Note that the client cost over iterations could not be the same. However, to make the problem more tractable, according to (9) we consider minimizing the upper-bound of the cost instead of the actual cost, similar to approach in [111].

4.2.2   Stackelberg Game-Based Solution

In this section, firstly, we present our motivation to realize the concept of federated learning by employing a crowdsourcing framework. We next advocate an incentive mechanism required to realize this setting of decentralized learning model with our proposed solution approach.

Incentive Mechanism: A Two-Stage Stackelberg Game Approach

The MEC server will allocate rewards to the participating clients to achieve optimal local accuracy in consideration for improving the communication efficiency of the system. That means the MEC server will plan to incentivize clients for maximizing its own benefit, i.e., an improved global model. Consequently, upon receiving the announced reward, any rational client will individually maximize their own profit. Such an interaction scenario can be realized with a Stackelberg game approach.

Specifically, we formulate our problem as a two-stage Stackelberg game between the MEC server (leader) and participating clients (followers). Under the crowdsourcing framework, the MEC server designs an incentive mechanism for participating clients to attain a local consensus accuracy level[6] on the local models while improving the performance of a centralized model. The MEC server cannot directly control the participating clients to maintain a local consensus accuracy level and requires an effective incentive plan to enroll clients for this setting.

Clients (Stage II) The MEC server has an advantage, being a leader with the first-move advantage influencing the followers for participation with a local consensus accuracy. It will at first announce a uniform reward rate[7] (e.g., a fair data package discount as $/accuracy level) $r > 0$ for the participating clients. Given r, at Stage II, a rational client k will try to improve the local model's accuracy for maximizing its net utility by training over the local data with global parameters. The proposed utility framework incorporates the cost involved while a client tries to maximize its own individual utility.

Client Utility Model We use a valuation function $v_k(\theta_k)$ to denote the model's effectiveness that explains the valuation of the client k when relative θ_k accuracy is attained for the local subproblem.

Assumption 1 The valuation function $v_k(\theta_k)$ is a linear, decreasing function with $\theta_k > 0$, i.e., $v_k(\theta_k) = (1 - \theta_k)$. Intuitively, for a smaller relative accuracy at the local subproblem, there will be an increase in the reward for the participating clients.

---

[6] It signifies the agreement among the participating clients on the quality of solution at the local subproblems for building a high-quality centralized learning model.

[7] Prominently, two kinds of pricing schemes exist at present following different design goals: uniform pricing and discriminatory or differentiated pricing [120]. The differentiated pricing scheme is more efficient, but also requires more information and higher complexity than the uniform pricing [121, 122]. Therefore, based upon offered motivations and benefits, our proposed crowdsourcing framework follows a platform-centric model to train a high-quality global model with low complexity, less information exchange by using the uniform pricing scheme.

Given $r > 0$, each participating client k's strategy is to maximize its own utility as follows:

$$\max_{0 \leq \theta_k \leq 1} \quad u_k(r, \theta_k) = r(1 - \theta_k) - C_k(\theta_k), \tag{4.16}$$

given cost $C_k(\theta_k)$ as (4.15). The feasible solution is always restricted to the value less than 1 (i.e., without loss of generality, for $\theta_k > 1$, it violates the participation assumption for the crowdsourcing framework). Therefore, problem (4.16) can be represented as

$$\max_{\theta_k > 0} \quad u_k(r, \theta_k) = r(1 - \theta_k) - C_k(\theta_k), \forall k \in \mathcal{K}. \tag{4.17}$$

Also, we have $C_k''(\theta_k) > 0$, which means $C_k(\theta_k)$ is a strictly convex function. Thus, there exists a unique solution $\theta_k^*(r), \forall k$.

MEC Server (Stage I) Knowing the response (strategy) of the participating clients, the MEC can evaluate an optimal reward rate r^* to maximize its utility. The utility $U(\cdot)$ of the MEC server can be defined in relation to the satisfaction measure achieved with the local consensus accuracy level.

MEC Server Utility Model We define $x(\epsilon)$ as the number of iterations required for an arbitrary algorithm to converge to some ϵ accuracy. We similarly define $I^g(\epsilon, \theta)$ as global iterations of the framework to reach a relative θ accuracy on the local subproblems.

From this perspective, we require an appropriate utility function $U(\cdot)$ as the satisfaction measure of the framework with respect to the number of iterations for achieving ϵ accuracy. In this regard, use the definition of the number of iterations for ϵ accuracy as

$$x(\epsilon) = \zeta \cdot \log\left(\frac{1}{\epsilon}\right).$$

Due to large values of iterations, we approximate $x(\epsilon)$ as a continuous value, and with the aforementioned relation, we choose $U(\cdot)$ as a strictly concave function of $x(\epsilon)$ for $\epsilon \in [0, 1]$, i.e., with the increase in $x(\epsilon)$, $U(\cdot)$ also increases. Thus, we propose $U(x(\epsilon))$ as the normalized utility function bounded within $[0, 1]$ as

$$U(x(\epsilon)) = 1 - 10^{-(ax(\epsilon)+b)}, \quad a \geq 0, b \leq 0, \tag{4.18}$$

which is strictly increasing with $x(\epsilon)$, and represents the satisfaction of MEC increase with respect to accuracy ϵ.

As for the global model, there exists an acceptable value of threshold accuracy measure correspondingly reflected by $x_{\min}(\epsilon)$. This suggests the possibility of near-zero utility for the MEC server for failing to attain such value.

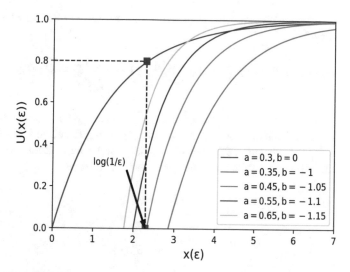

Fig. 4.6 MEC utility $U(\cdot)$ as a function of ϵ with different parameter values of a, b

Figure 4.6 depicts our proposed utility function, a concave function of $x(\epsilon)$ with parameters a and b that reflect the required behavior of the utility function defined in (4.18). In Fig. 4.6, we can observe that a larger value of a means smaller iterations requirement and larger values of b introduces flat curves suggesting more flexibility in accuracy. So we can analyze the impact of parameters a and b in (4.18), and set them to model the utility function for the MEC server as per the design requirements of the learning framework. Furthermore, in our setting, $I^g(\epsilon, \theta)$ can be written as

$$I^g(\epsilon, \theta) = \frac{x(\epsilon)}{1 - \theta} \leq \delta. \tag{4.19}$$

(4.19) explains the efficiency paradigm of the proposed framework in terms of time required for the convergence to some accuracy ϵ. If $\tau^l(\theta)$ is the time per iteration to reach a relative θ accuracy at a local subproblem and $T(\theta)$ is the communication time required during a single iteration for any arbitrary algorithm, then we can analyze the result in (4.19) with the efficiency of the global model as

$$I^g(\epsilon, \theta) \cdot (T(\theta) + \tau^l(\theta)). \tag{4.20}$$

Because the cost of communication is proportional to the speed and energy consumption in a distributed scenario [123], the bound defined in (4.19) explains the efficiency in terms of the MEC server's resource restriction for attaining ϵ accuracy. In this regard, the corresponding analysis of (4.20) is presented in the upcoming sub-section with several case studies.

The utility of the MEC server can therefore be defined for the set of measured best responses $\boldsymbol{\theta}^*$ as

$$\mathcal{U}(x(\epsilon), r|\boldsymbol{\theta}^*) = \beta \left(1 - 10^{-(ax(\epsilon)+b)}\right) - r \sum_{k \in \mathcal{K}} (1 - \theta_k^*(r)),$$

where $\beta > 0$ is the system parameter,[8] and $r \sum_{k \in \mathcal{K}} (1 - \theta_k^*(r))$ is the cost spent for incentivizing participating clients in the crowdsourcing framework for federated learning. So, for the measured $\boldsymbol{\theta}^*$ from the participating clients at the MEC server, the utility maximization problem can be formulated as follows:

$$\max_{r \geq 0, x(\epsilon)} \quad \mathcal{U}(x(\epsilon), r|\boldsymbol{\theta}^*), \tag{4.21}$$

$$\text{s.t.} \quad \frac{x(\epsilon)}{1 - \max_k \theta_k^*(r)} \leq \delta. \tag{4.22}$$

In constraint (4.22), $\max_k \theta_k^*(r)$ characterizes the worst-case response for the server-side utility maximization problem with the bound on permissible global iterations. Note that MEC adapts admission control strategy (discussed in Sect. 4.2.3) to improve the number of participants for maximizing its utility. In fact, MEC has to increase the reward rate to maintain the minimum number of participation (at least two) to realize the distributed optimization setting in federated learning. In addition to this, the framework may suffer from slower convergence due to less participation. Thus, MEC will avoid deliberately dropping the clients to achieve a faster consensus with (4.22).

Furthermore, using the relationship defined in (4.19) between $x(\epsilon)$ and relative θ accuracy for the subproblem, we can analyze the impact of responses $\boldsymbol{\theta}$ on MEC server's utility in a federated learning setting with the constraint (4.11). To be more specific about this relation, we can observe that with the increased value of $(1 - \theta)$, i.e., lower relative accuracy (high local accuracy), the MEC server can attain better utility due to the corresponding increment in the value of $x(\epsilon)$. Note that in the client cost problem, $x(\epsilon)$ is treated as a constant provided by the MEC problem, and can be ignored for solving (4.16).

Lemma 1 *The optimal solution* $x^*(\epsilon)$ *for (4.21) can be derived as* $\delta(1 - \max_k \theta_k^*(r))$.

Proof See Appendix.

---

[8] Note that $\beta > 0$ characterizes a linear scaling metric to the utility function which can be set arbitrarily and will not alter our evaluation. Equivalently, it can be understood as the MEC server's physical resource consignments for the federated learning that reflects the satisfaction measure of the framework.

Therefore, for the given $\theta^*(r)$, we can formalize (4.21) as

$$\max_{r \geq 0} \quad \beta \left(1 - 10^{-(ax^*(\epsilon)+b)}\right) - r \sum_{k \in \mathcal{K}} (1 - \theta_k^*(r)). \tag{4.23}$$

Stackelberg Equilibrium With a solution to MEC server's utility maximization problem, r^* we have the following definition.

Definition 1 For any values of r, and θ, (r^*, θ^*) is a Stackelberg equilibrium if it satisfies the following conditions:

$$\mathcal{U}(r^*, \theta^*) \geq \mathcal{U}(r, \theta^*), \tag{4.24}$$

$$u_k(\theta_k^*, r^*) \geq u_k(\theta_k, r^*), \quad \forall k. \tag{4.25}$$

Next, we employ the backward-induction method to analyze the Stackelberg equilibria: the Stage-II problem is solved at first to obtain θ^*, which is then used for solving the Stage-I problem to obtain r^*.

Stackelberg Equilibrium: Algorithm and Solution Approach

Intuitively, from (4.19), we see that the server can evaluate the maximum value of $x(\epsilon)$ required for attaining accuracy ϵ for the centralized model while maintaining relative accuracy θ_{th} amongst the participating clients. Here, θ_{th} is a consensus on a maximum local accuracy level amongst participating clients, i.e., the local subproblems will maintain at least θ_{th} relative accuracy. So, with the measured responses θ from the participating clients, the server can design a proper incentive plan to improve the global model while maintaining the worst-case relative accuracy $\max_k \theta_k^*$ as θ_{th} for the local model.

Since the threshold accuracy θ_{th} can be adjusted by the MEC server for each round of solution, each participating client will maintain a response towards the maximum local consensus accuracy θ_{th}. This formalizes the client's selection criteria [see Remark 1.] which is sufficient enough for the MEC server to maintain the accuracy ϵ. We also have the lower bound related with the value of $x_{\min}(\epsilon)$ for equivalent accuracy $\epsilon_{\max}$ while dealing with the client's responses θ, i.e.,

$$\log\left(\frac{1}{\epsilon_{\max}}\right) \leq \frac{x(\epsilon)}{(1 - \theta_{\text{th}})} \leq \delta_{\max}. \tag{4.26}$$

where $\delta_{\max}$ is the maximum permissible upper bound to the global iterations.

As explained before and with (4.26), the value of θ_{th} can be varied (lowered) by MEC server to improve the overall performance of the system. For a worst case scenario, where the offered reward r for the client k is insufficient to motivate it for participation with improved local relative accuracy, we might have $\max_k \theta_k^*(r) = 1$, i.e., $\theta_{\text{th}} = 1$, no participation.

Lemma 2 *For a given reward rate r, and T_k which is determined based upon the channel conditions (4.14), we have the unique solution $\theta_k^*(r)$ for the participating client satisfying following relation:*

$$g_k(r) = \log(e^{1/\theta_k^*(r)}\theta_k^*(r)), \forall k \in \mathcal{K}, \tag{4.27}$$

for $g_k(r) \geq 1$, where,

$$g_k(r) = \left[\frac{r + v_k T_k}{(1 - v_k)\gamma_k} - 1\right].$$

Proof Because $C_k''(\theta_k) > 0$ for $\theta_k > 0$, (4.17) is a strictly convex function resulted as a linear plus convex structure. Therefore, by the first-order condition, (4.17) can be deduced as

$$\frac{\partial u_k(r, \theta_k)}{\partial \theta_k} = 0$$

$$\Leftrightarrow \frac{1}{\theta_k} - \log\left(\frac{1}{\theta_k}\right) = \left[\frac{r + v_k T_k}{(1 - v_k)\gamma_k} - 1\right], \tag{4.28}$$

$$\Leftrightarrow \log(e^{1/\theta_k}\theta_k) = g_k(r).$$

We observe that Lemma 2 is a direct consequence of the solution structure derived in (4.28). Hence, we conclude the proof.

From Lemma 2, we have some observations with the definition of $g_k(r)$ for the response of the participating clients. First, we can show that θ_k^* is larger for the poor channel condition on a given reward rate. Second, in such scenario, with the increase in reward rate, say for $g_k(r) > 2$ the participating clients will iterate more during their computation phase resulting in lower θ_k^*. This will reduce the number of global iterations to attain an accuracy level for the global problem.

We can therefore characterize the participating client k's best response under the proposed framework as

$$\theta_k^*(r) = \min\left\{\hat{\theta}_k(r) \mid_{g_k(r)=\log(e^{1/\hat{\theta}_k(r)}\hat{\theta}_k(r))}, \theta_{\text{th}}\right\}, \forall k. \tag{4.29}$$

(4.29) represents the best response strategy for the participating client k under our proposed framework. Intuitively, exploring the logarithmic structure in (4.27), we observe that the increase in incentive r will motivate participating clients to increase their efforts for local iteration in one global iteration. This is reflected by a better response, i.e., a lower relative accuracy (high local accuracy) during each round of communication with the MEC server.

Figure 4.7 illustrates such strategic responses of the participating clients over an offered reward for a given configuration. In this scenario, to elaborate the best

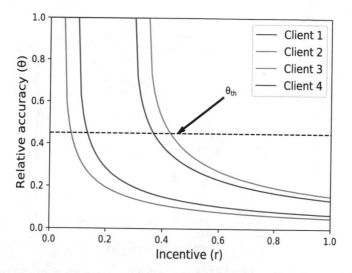

Fig. 4.7 An illustration showing participating clients response over the offered reward rate

response strategy as characterized in (4.29), we have considered four participating clients with different preferences (e.g., Client 3 being the most reluctant participant). We observe that Client 3 seeks more incentive r to maintain a comparable accuracy level as Client 1. Further, we consider the tradeoff between communication cost and the computation cost as discussed with the relation in (4.15). These costs are complementary in relation by v_k, and for each client k their preferences upon these costs are also different. For instance, the higher value of v_k for client k emphasizes the increased number of communication with the MEC server to improve the local relative accuracy θ_k.

In Figs. 4.8, 4.9, and 4.10, we briefly present the solution analysis to (4.27) with the impact of channel condition (we define it as communication adversity) on the local relative accuracy for a constant reward. For this, in Fig. 4.8 we consider a participating client with the fixed offered reward setting r from uniformly distributed values of 0.1–5. We use normalized T_k parameter for a client k to illustrate the response analysis scenario. In Figs. 4.9 and 4.10, T_k is uniformly distributed on [0.1, 1], and v_k is set at 0.6. Intuitively, as in Fig. 4.8, the increase in communication time T_k for a fixed reward r will influence participating clients to iterate more locally for improving local accuracy than to rely upon the global model, which will minimize their total cost. Under this scenario, we observe the increase in communication cost with the increase in communication time T_k. Thus, the clients will iterate more locally. However, the trend is significantly affected by normalized weights v_k, as observed in Figs. 4.9 and 4.10. For a larger value of T_k (poor channel condition) as in the case of Fig. 4.10, increasing the value of v_k, i.e., clients with more preference on the communication cost in the total cost model results to higher local iterations for solving local subproblems, as reflected by the better local accuracy, unlike in Fig. 4.9. In both cases we observe the decrease in

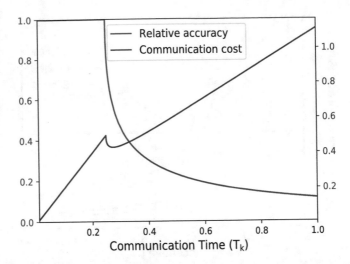

Fig. 4.8 Solution Analysis (4.27) (Left Y-axis: Relative accuracy, Right Y-axis: Communication cost): impact of communication adversity on local relative accuracy for a constant reward

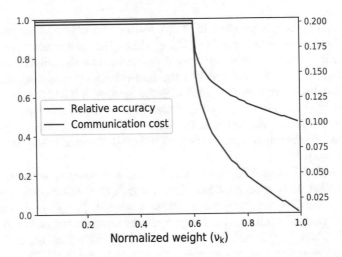

Fig. 4.9 Solution Analysis (4.27) (Left Y-axis: Relative accuracy, Right Y-axis: Communication cost): normalized weight versus relative accuracy for a fair data rate (quality communication channel)

communication cost upon participation. However, in Fig. 4.10 the communication cost is higher because of an expensive data rate. Therefore, for a given r, client k can adjust its weight metrics accordingly to improve the response θ_k.

In Figs. 4.11, 4.12, and 4.13, we explore such behaviors of the participating clients through the heatmap plot. To explain better, we define three categories of participating clients based upon the value of normalized weights $v_k, \forall k$, which are their individual preferences upon the computation cost and the communication

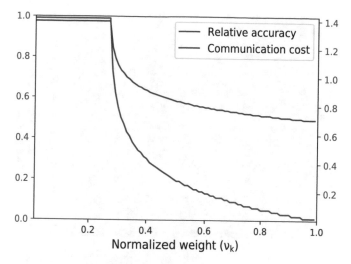

Fig. 4.10 Solution Analysis (4.27) (Left Y-axis: Relative accuracy, Right Y-axis: Communication cost): normalized weight versus relative accuracy for an expensive data rate

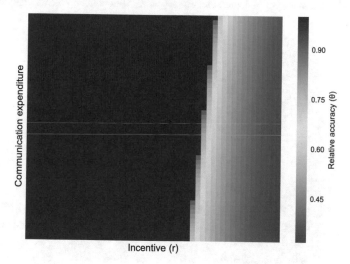

Fig. 4.11 Case Study: impact of communication cost and offered reward rate r for normalized weight (preferences), reluctant clients i.e., $v_k = 0.1$. X-axis shows the increase in incentive (r) value from left-to-right, and the y-axis defines the increase in value of communication expenditure (top-to-bottom)

cost for the convergence of the learning framework. (i) *Reluctant* clients with a lower v_k consume more reward to improve local accuracy, even though the value of T_k is larger (expensive), as observed in Fig. 4.11. (ii) *Sensitive* clients are more susceptible towards the channel quality with larger v_k, and iterates more locally within a round of communication to the MEC server for improving local accuracy,

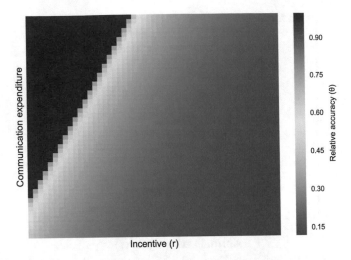

Fig. 4.12 Case Study: impact of communication cost and offered reward rate r for normalized weight (preferences), rational clients, $v_k = 0.7$. X-axis shows the increase in incentive (r) value from left-to-right, and the y-axis defines the increase in value of communication expenditure (top-to-bottom)

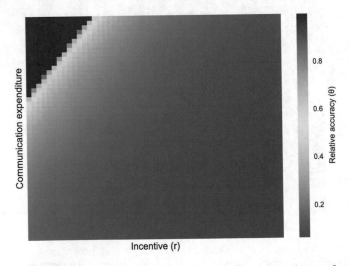

Fig. 4.13 Case Study: impact of communication cost and offered reward rate r for normalized weight (preferences), sensitive clients, $v_k = 0.7$. X-axis shows the increase in incentive (r) value from left-to-right, and the y-axis defines the increase in value of communication expenditure (top-to-bottom)

as observed in Fig. 4.13. (iii) *Rational* clients, as referred in Fig. 4.12 tend to balance these extreme preferences (say $v_k = 0.5$ for client k), which in fact would be unrealistic to expect all the time due to heterogeneity in participating client's resources.

Algorithm 5 MEC server's utility maximization

1: Sort clients as with $\hat{r}_1 < \hat{r}_2 < \ldots < \hat{r}_K$
2: $\mathcal{R} = \{\}, \mathcal{A} = \mathcal{K}, j = K$
3: **while** $j > 0$ **do**
4:     Obtain the solutions r_j to the following problem:

$$\max_{r \geq \hat{r}_1} \beta \left(1 - 10^{-(ax^*(\epsilon)+b)}\right) - r \sum_{k \in \mathcal{A}} (1 - \theta_k^*(r))$$

5:     **if** $r_j > \hat{r}_j$, **then** $\mathcal{R} = \mathcal{R} \cup \{r_j\}$;
6:     **end if**
7:     $\mathcal{A} = \mathcal{A} \backslash j$;
8:     $j = j - 1$;
9: **end while**
10: Return $r_j \in \mathcal{R}$ with highest optimal values in problem (4).

To solve (4.23) efficiently, with (4.29) $\theta_k^*(r) = \min \left\{ \hat{\theta}_k(r) \mid_{g_k(r)=\log(e^{1/\hat{\theta}_k(r)}\hat{\theta}_k(r))}, \theta_{\text{th}} \right\}$, $\forall k$, we introduce a new variable z_k in relation with consensus on local relative accuracy θ_{th},

$$z_k = \begin{cases} 1, & \text{if } r > \hat{r}_k; \\ 0, & \text{otherwise,} \end{cases} \tag{4.30}$$

where

$$\hat{r}_k = \left[g_k^{-1}(\log(e^{1/\theta_{\text{th}}}\theta_{\text{th}})) \right]$$

is the minimum incentive value required obtained from (4.29) to attain the local consensus accuracy θ_{th} at client k for the defined parameters ν_k and T_k.

This means, $\theta_k(r) < \theta_{\text{th}}$ when $z_k = 1$, and $\theta_{\text{th}} \leq \theta_k(r) < 1$ when $z_k = 0$. MEC server can use this setting to drop the participants with poor accuracy. As discussed before, for the worst case scenario we consider $\theta_{\text{th}} = 1$.

Therefore, the utility maximization problem can be equivalently written as

$$\max_{r, \{z_k\}_{k \in \mathcal{K}}} \beta \left(1 - 10^{-(ax^*(\epsilon)+b)}\right) - r \sum_{k \in K} z_k \cdot (1 - \theta_k^*(r)), \tag{4.31}$$

$$\text{s.t.} \quad r \geq 0, \tag{4.32}$$

$$z_k \in \{0, 1\}, \forall k. \tag{4.33}$$

The problem (4.31) is a mixed-Boolean programming, which may require exponential-complexity effort (i.e., 2^K configuration of $\{z_k\}_{k \in \mathcal{K}}$) to solve by the exhaustive search. To solve this problem with linear complexity, we refer to the solution approach as in Algorithm 5.

The utility maximization problem at MEC server can be reformulated as a constraint optimization problem (4.34–4.35) assuming a fixed configuration of $\{z_k = 1\}_{k \in \mathcal{K}}$ as

$$\max_{r \geq 0} \quad \beta \left(1 - 10^{-(ax^*(\epsilon)+b)}\right), \tag{4.34}$$

$$\text{s.t.} \quad r \sum_{k \in K}(1 - \theta_k^*(r)) \leq B, \tag{4.35}$$

where (4.35) is budget constraint for the problem. The second-order derivative of function $r(1 - \theta_k^*(r))$ in (4.35) is $\frac{2\gamma_k(1-v_k)v_k T_k}{(r+v_k T_k)^3} > 0$, i.e., the problem (4.34) is a convex problem and can be solved similarly with Algorithm 5 (line 4–5).

Proposition 1 *Algorithm 2 can solve the Stage-I equivalent problem (4.23) with linear complexity.*

Proof As the clients are sorted in the order of increasing $\hat{r}_k$ (line 1), for the sufficient condition $r > \hat{r}_k$ resulting $z_k = 1$, the MEC's utility maximization problem reduces to a single-variable problem that can be solved using popular numerical methods.

Remark 1 Algorithm 2 can maintain consensus accuracy by formalizing the clients selection criteria. This is because from (4.30), $z_k = 1$ for $\theta_k(r) < \theta_{\text{th}}$, and $z_k = 0$ for $\theta_{\text{th}} \leq \theta_k(r) < 1$. Thus, MEC server uses this setting to drop the participants with $\theta_k(r) > \theta_k^*(r) = \theta_{\text{th}}$.

Theorem 1 *The Stackelberg equilibria of the crowdsourcing framework are the set of pairs $\{r^*, \boldsymbol{\theta}^*\}$.*

Proof For any given $\boldsymbol{\theta}$, it is obvious that $\mathcal{U}(r^*, \boldsymbol{\theta}) \geq \mathcal{U}(r, \boldsymbol{\theta}), \forall r$ since r^* is the solution to the Stage-I problem. Thus, we have $\mathcal{U}(r^*, \boldsymbol{\theta}^*) \geq \mathcal{U}(r, \boldsymbol{\theta}^*)$. In the similar way, for any given value of r and $\forall k$, we have $u_k(r, \theta_k^*) \geq u_k(r, \theta_k), \forall \theta_k$. Hence, $u_k(r^*, \theta_k^*) \geq u_k(r^*, \theta_k)$. Combining these facts, we conclude the proof being based upon the definitions of (4.24) and (4.25).

4.2.3   Simulations

In this section, we present numerical simulations to illustrate our results. We consider the learning setting for a strongly convex model such as logistic regression, as discussed in Sect. 4.2.1, to characterize and demonstrate the efficacy of the proposed framework. First, we will show the optimal solution of Algorithm 5 (Algorithm 5) and conduct a comparison of its performance with two baselines. The first one, named OPT, is the optimal solution of problem (4.23) with an exhaustive search for the optimal response $\boldsymbol{\theta}^*$. The second one is called Baseline that considers the worst response amongst the participating clients to attain local consensus θ_{th}

accuracy with an offered price. This is an inefficient scheme but still enables us to attain feasible solutions. Finally, we analyze the system performance by varying different parameters and conduct a comparison of the incentive mechanism with the baseline and their corresponding utilities. In our analysis, the smaller values of local consensus are of specific interest as they reflect the effectiveness of federated learning.

1. *Settings:* For an illustrative scenario, we fix the number of participating clients to 4. We consider the system parameter $\beta = 10$, and the upper bound to the number of global iterations $\delta = 10$, which characterizes the permissible rounds of communication to ensure global ϵ accuracy. The MEC's utility $U(x(\epsilon)) = 1 - 10^{-(ax(\epsilon)+b)}$ model is defined with parameters $a = 0.3$, and $b = 0$. For each client k, we consider normalized weight v_k is uniformly distributed on [0.1,0.5], which can provide an insight on the system's efficacy as presented in Figs. 4.11, 4.12, and 4.13. We characterize the interaction between the MEC server and the participating clients under homogeneous channel condition, and use the normalized value of T_k for all participating clients.

2. *Reward rate:* In Fig. 4.14 we increase the value of local consensus accuracy θ_{th} from 0.2 to 0.6. When the accuracy level is improved (from 0.4 to 0.2), we observe a significant increase in the reward rate. These results are consistent with the analysis in section "Stackelberg Equilibrium: Algorithm and Solution Approach". The reason is that cost for attaining a higher local accuracy level

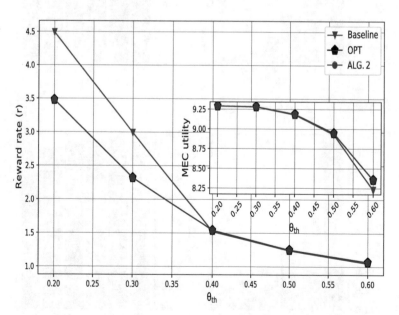

Fig. 4.14 Comparison of (**a**) Reward rate and (**b**) MEC utility under three schemes for different values of threshold θ_{th} accuracy

requires more local iterations, and thus the participating clients exert more incentive to compensate for their costs.

We also show that the reward variation is prominent for lower values of θ_{th}, and observe that scheme Algorithm 2 and OPT achieve the same performance, while Baseline is not as efficient as others. Here, we can observe up to 22% gain in the offered reward against the Baseline by other two schemes. In Fig. 4.14b, we see the corresponding MEC utilities for the offered reward that complements the competence of the proposed Algorithm 2. We see, the trend of utility against the offered reward goes along with our analysis.

3. *Parametric choice:* In Figs. 4.15 and 4.16 we show the impact of parametric choice adopted by the participating client k to solve the local subproblem [124], which is characterized by γ_k. In Fig. 4.15, we see a lower offered reward for the improved local accuracy level for the participating clients adapting same

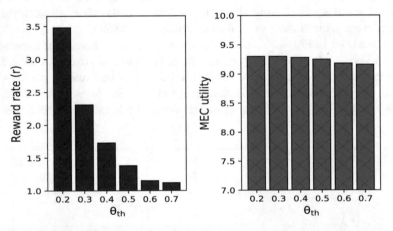

Fig. 4.15 For $|\mathcal{K}| = 4$, $a = 0.3$, $b = 0$, $\gamma_k = 1$, $\forall k$

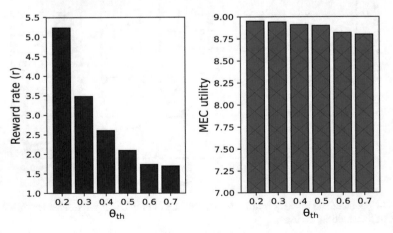

Fig. 4.16 For $|\mathcal{K}| = 4$, $a = 0.3$, $b = 0$, and $\gamma_k \sim U[1, 5]$

Table 4.2 Offered reward rate comparison with randomized γ effect for different (a, b) setting

Threshold accuracy θ_{th}	Baseline r	Algorithm 2 $(0.3, -1)$	Algorithm 2 $(0.35, -1)$	Algorithm 2 $0.65, -1)$
0.2	18	5.22	5.22	5.22
0.3	12	3.48	3.48	3.48
0.4	8.99	2.602	2.6	2.61
0.5	7.19	2.79	4.3	2.2
0.6	5.99	2.38	2.87	2.1
0.7	5.13	2.84	3.17	1.9

Table 4.3 Utility comparison with randomized γ effect for different (a, b) setting

Threshold accuracy θ_{th}	Algorithm 2 $(0.3, -1)$	Algorithm 2 $(0.35, -1)$	Algorithm 2 $(0.65, -1)$
0.2	8.55	8.79	8.96
0.3	8.41	8.60	8.95
0.4	8.33	8.58	8.94
0.5	8.2	8.73	8.91
0.6	8.18	8.4	8.91
0.7	7.8	8.51	8.86

parameters (algorithms) for solving the local subproblem, in contrast to Fig. 4.16 with the uniformly distributed γ_k on [1,5] to achieve the competitive utility.

4. *Comparisons:* In Tables 4.2 and 4.3, we see the effect of randomized parameter γ_k for different configuration of MEC utility model $U(\cdot)$ defined by (a, b). For the smaller values of θ_{th}, which captures the competence of the proposed mechanism, we observe that the choice of (a, b) provides a consistent offered reward for improved utility from $(0.35, -1)$ to $(0.65, -1)$, which follows our analysis in section "Incentive Mechanism: A Two-Stage Stackelberg Game Approach". For larger values of θ_{th}, we also see the similar trend in MEC utility. For a randomized setting, we observe up to 71% gain in offered reward against the Baseline, which validates our proposal's efficacy aiding federated learning.

Our earlier discussion in Sect. 4.2.2 and simulation results explain the significance of choosing a local θ_{th} accuracy to build a global model that maximizes the utility of the MEC server. In this regard, at first, the MEC server evokes admission control to determine θ_{th} and the final model is learned later. This means, with the number of expected clients, it is crucial to appropriately select a proper prior value of θ_{th} that corresponds to the participating client's selection criteria for training a specific learning model. Note that, in each communication round of synchronous aggregation at the MEC server, the quality of local solution benefits to evaluate the performance at the local subproblem. In this section, we will discuss about the probabilistic model employed by the MEC server to determine the value of the consensus θ_{th} accuracy.

We consider the local θ accuracy for the participating clients is an i.i.d and uniformly distributed random variable over the range $[\theta_{min}, \theta_{max}]$, then the PDF of the responses can be defined as $f_\theta(\theta) = \frac{1}{\theta_{max} - \theta_{min}}$. Let us consider a sequence of discrete time slots $t \in \{1, 2, \ldots\}$, where the MEC server updates its configuration for improving the accuracy of the system. Following our earlier definitions, at time slot t, the number of participating clients in the crowdsourcing framework for federated learning is $|\mathcal{K}(t)|$, or simply K. We restrict the clients with the accuracy measure $\boldsymbol{\theta}(t) \geq \theta_{max}$. For K number of participation requests, the total number of accepted responses $N(t)$ is defined as $N(t) = K \cdot F_{\boldsymbol{\theta}(t)}(\theta) = K \cdot P[\boldsymbol{\theta}(t) \leq \theta]$. We have $N(t) = K \cdot \left[\frac{\theta(t) - \theta_{min}}{\theta_{max} - \theta_{min}}\right]$. At each time t, the MEC server chooses $\boldsymbol{\theta}(t)$ as the threshold accuracy θ_{th} that maximizes the sum of its utility as defined in (4.18) for the defined parameters $a \geq 0, b \leq 0$ and the total participation, $\beta\left(1 - 10^{-(ax(\epsilon)+b)}\right) + (1 - \theta) \cdot N(t)$, subject to the constraint that the response lies between the minimum and maximum accuracy measure ($\theta_{min} \leq \boldsymbol{\theta}(t) \leq \theta_{max}$). Using the definitions in (4.19), for $\beta > 0$, the MEC server maximizes its utility for the number of participation with θ accuracy as

$$\max_{\theta(t)} \quad \beta\left(1 - 10^{-(a \cdot \delta(1-\theta(t))+b)}\right) + (1 - \theta(t)) \cdot N(t),$$

$$\text{s.t.} \quad \theta_{min} \leq \theta(t) \leq \theta_{max}. \tag{4.36}$$

The Lagrangian of the problem (4.36) is as follows:

$$\mathcal{L}(\theta(t), \lambda, \mu) = \beta\left(1 - 10^{-(a \cdot \delta(1-\theta(t))+b)}\right) + (1 - \theta(t)) \cdot$$
$$\left[\frac{\theta(t) - \theta_{min}}{\theta_{max} - \theta_{min}}\right] + \lambda(\theta(t) - \theta_{min})$$
$$+ \mu(\theta_{max} - \theta(t)), \tag{4.37}$$

where $\lambda \geq 0$ and $\mu \geq 0$ are dual variables. Problem (4.36) is a convex problem whose optimal primal and dual variables can be characterized using the Karush-Khun-Tucker (KKT) conditions [116] as

$$\frac{\partial \mathcal{L}}{\partial \theta(t)} = \ln(10) \cdot (\beta \delta a) \cdot 10^{-(a \cdot \delta(1-\theta^*(t))+b)}$$

$$-K \cdot \left[\frac{2\theta(t) - \theta_{min}}{\theta_{max} - \theta_{min}}\right] + \lambda - \mu = 0, \tag{4.38}$$

$$\lambda(\theta(t) - \theta_{min}) = 0, \tag{4.39}$$

$$\nu(\theta_{max} - \theta(t)) = 0. \tag{4.40}$$

Following the complementary slackness criterion, we have

$$\lambda^*(\theta^*(t) - \theta_{\min}) = 0, \; \mu^*(\theta_{\max} - \theta^*(t)) = 0, \; \lambda^* \geq 0, \; \mu^* \geq 0. \tag{4.41}$$

Therefore, from (4.41), we solve (4.36) with the KKT conditions assuming that $\theta^*(t) < \theta_{\max}$ as an admission control strategy, and find the optimal $\theta^*(t)$ that satisfies the following relation

$$K = \frac{\ln(10) \cdot (\beta\delta a) \cdot 10^{-(a\cdot\delta(1-\theta^*(t))+b)} \cdot (\theta_{\min} - \theta_{\max})}{1 - 2\theta^*(t) + \theta_{\min}}. \tag{4.42}$$

(4.42) can be rearranged as

$$f(\theta^*(t)) = \ln(10) \cdot (\beta\delta a) \cdot 10^{-(a\cdot\delta(1-\theta^*(t))+b)}$$
$$+ K \cdot \left[\frac{1 - 2\theta^*(t) + \theta_{\min}}{\theta_{\max} - \theta_{\min}} \right] = 0. \tag{4.43}$$

To obtain the value of $\theta^*(t)$ we will use *Netwon-Raphson method* [125] employing an appropriate initial guess that manifests the quadratic convergence of the solution. We choose $\theta_0^*(t) = E(\boldsymbol{\theta}(t)) = \frac{\theta_{\max}+\theta_{\min}}{2}$ as an initial guess for finding $\theta^*(t)$ which follows the PDF $f_{\boldsymbol{\theta}}(\theta) \sim U[\theta_{\min}, \theta_{\max}]$. Then the solution method is an iterative approach as follows:

$$\theta_{i+1}^*(t) = \theta_i^*(t) - \frac{f(\theta_i^*(t))}{\beta\delta^2 a^2 \cdot \ln^2(10) \cdot 10^{-(a\cdot\delta(1-\theta_i^*(t))+b)}}. \tag{4.44}$$

Numerical Analysis: In Figs. 4.17 and 4.18, we vary the number of participating clients up to 50 with different values of δ. The response of the clients is set to follow a uniform distribution on $[0.1, 0.9]$ for the ease of representation. In Fig. 4.17, for the model parameters (a,b) as $(0.35, -1)$, we see θ_{th} increases with the increase in the number of participating clients for all values of δ. It is intuitive and goes along with our earlier analysis that for the small number of participating clients, the smaller θ_{th} captures the efficacy of our proposed framework. Because it is an iterative process, the evolution of θ_{th} over the rounds of communication will be reflected in the framework design. Subsequently, the larger upper bound δ exhibits the similar impact on setting θ_{th}, where smaller δ imposes strict local accuracy level to attain high-quality centralized model. Also due to the same reason, in Fig. 4.18, we see θ_{th} is increasing for the increase in the number of participating clients, however, with the lower value. It is because of the choice of parameters (a, b) as explained in section "Incentive Mechanism: A Two-Stage Stackelberg Game Approach". So the value of θ_{th} is lower in Fig. 4.18.

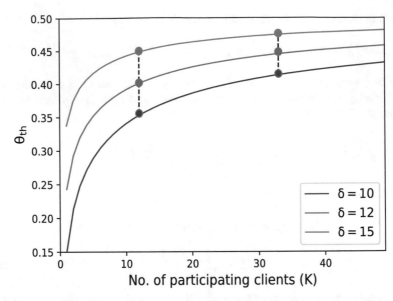

Fig. 4.17 Variation of local θ_{th} accuracy for different values of δ given the density function, $f_{\boldsymbol{\theta}}(\theta) \sim U[0.1, 0.9]$, $|\mathcal{K}| = [0, 50]$, for $a = 0.35, b = -1$

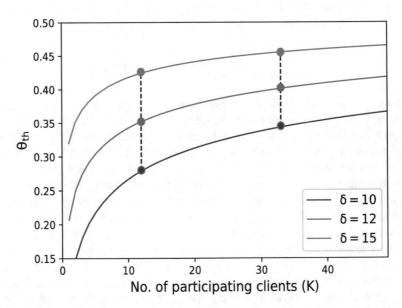

Fig. 4.18 Variation of local θ_{th} accuracy for different values of δ given the density function, $f_{\boldsymbol{\theta}}(\theta) \sim U[0.1, 0.9]$, $|\mathcal{K}| = [0, 50]$, for $a = 0.45, b = -1.05$

4.3   Auction Theory-Enabled Incentive Mechanism

In a typical wireless system, there are a variety of players, such as end-users, mobile network operators, and cloud server providers, among others. These players interact with each other to maximize their own benefits [126]. The benefits can be high data rate, load balancing, latency minimization, overall system utility maximization, energy efficiency, and profit maximization. However, enabling interaction among various players of wireless systems requires some effective business models. One can use auction theory to enable efficient interaction among the players of wireless systems. Auction theory enables people how to act in an auction market and systematically investigate the auction markets [127]. For instance, a mobile network operator can sell the spectrum resources to end-users for maximizing its profit. On the other hand, end-users want to increase their own benefits (e.g., data rate). To model this interaction between a mobile network operator and a set of users, one can use auction theory where users act as buyers (i.e, bidders) and mobile network operator as a seller. The description of three main players used in auction theory are as follows [128]:

- **Bidder:** In auction theory, a bidder is an entity that wants to buy commodities from the seller. An example of a bidder in a wireless system is the end-user.
- **Seller:** A seller denotes the owner of all commodities (e.g., radio resources) of a wireless system.
- **Auctioneer:** It refers to the intermediate agent that helps the bidder and seller in performing auctions. For instance, a base station in a wireless system can be considered as an auctioneer between the mobile network operator and end-users. Furthermore, a mobile network operator itself can also act as an auctioneer. Therefore, depending on the scenario one can choose an auctioneer.
- **Commodity:** Commodities of a wireless system represent the resources (e.g., spectrum, edge computing resource) that are traded between the sellers and bidders.

In this section, we provide an incentive mechanism design for federated learning using auction theory. In spite of the many benefits of federated learning, there are remaining two key challenges of having an efficient federated learning framework. The first challenge is the economic challenge. Data samples per mobile device are small to train a high-quality learning model so a large number of mobile users are needed to ensure cooperation. In addition, the mobile users who join the learning process are independent and uncontrollable. Here, mobile users may not be willing to participate in the learning due to the energy cost incurred by model training. In other words, the base station (BS), which generates the global model, has to stimulate the mobile users for participation. The second challenge is the technical challenge. On the one hand, we need users to collectively provide a large number of data samples to enable federated learning without sharing their private data. On the other hand, we need to protect the model from imperfect updates. The global loss minimization problem should enable (a) proper assessment of the quality of the local solution to improve personalization and fairness amongst the participating clients while training a global model, (b) effective decoupling of the local solvers, thereby balancing communication and computation in the distributed setting. Moreover, we need to consider wireless resource limitations (such as time, antenna number, and bandwidth)affecting the performance of federated learning. Besides, the limited energy of wireless devices is a crucial challenge for deploying federated learning. Indeed, it is necessary to optimize the energy efficiency for federated learning implementation because of these resource constraints.

To deal with the above challenges, we model the federated learning service between the BS and mobile users as an auction game in which the BS is a buyer and mobile users are sellers. In particular, the BS first initiates and announces a federated learning task. When each mobile user receives the federated learning task information, they decide the amount of resources required to participate in the model training. After that, each mobile user submits a bid, which includes the required amount of resource, local accuracy, and the corresponding energy cost, to the BS. Moreover, the BS plays the role of the auctioneer to decide the winners among mobile users as well as clear payment for the winning mobile users. In addition, the auction used in this work is a type of combinational auction [56, 129] since each mobile user can bid for combinations of resources. However, the proposed auction mechanism allows mobile users sharing the resources at the BS, which is different from the conventional combinatorial auction. The proposed mechanism directly determines the trading rules between the buyer (BS) and sellers (mobile users) and motivates the mobile users to participate in the model training. Compared with other incentive mechanism approaches (e.g., contract theory) in which the service market is a monopoly market, where mobile users can only decide whether or not to accept the contracts, the proposed auction enables mobile users to bids on any combinations of resources. Moreover, the proposed auction mechanism can simultaneously provide truthfulness and individual rationality. An auction mechanism is truthful if a bidder's utility does not increase when that bidder makes other bidding strategies, rather than the true value. Revealing the true value is a dominant strategy for each participating user regardless of what strategies other users use[130]. An absent-truthfulness auction mechanism could leave the door to possible market manipulation and produce inferior results [131]. Additionally, if the value of any bidder is non-negative, an auction process will ensure individual rationality. The contributions of this section are summarized as follows:

- We propose an auction framework for the wireless federated learning services market. *Then, we present the bidding cost in every user's bid submitted to the BS.* From the mobile users' perspective, each mobile user makes optimal decisions on the amount of resources and local accuracy to minimize weighted sum of completion time and energy costs while the delay requirement for federated learning is satisfied. To solve the cost decision problem, a low-complexity iterative algorithm is proposed.

- From the perspective of the BS, we formulate the winner selection problem in the auction game as the social welfare maximization problem, which is an NP-hard problem considering the limitation of the wireless resource. We propose a primal-dual greedy algorithm to deal with the NP-hard problem of selecting the winning users and critical value-based payment. We also proved that the proposed auction mechanism is truthful, individually rational, and computationally efficient.

- Finally, we carry out the numerical study to show that a proposed auction mechanism can guarantee the approximation factor of the integrality to the maximal welfare that is derived by the optimal solution and outperforms compared with baseline.

4.3.1   System Model

Preliminary of Federated Learning

Consider a cellular network in which one BS and a set $\mathcal{N}$ of N users cooperatively perform a federated learning algorithm for model learning, as shown in Fig. 4.19. A summary of all notations used is given in Table 4.4. Each user n has s_n local data samples. Each data set $s_n = \{a_{nk}, b_{nk,1 \leq k \leq s_n}\}$ where a_{nk} is an input and b_{nk}

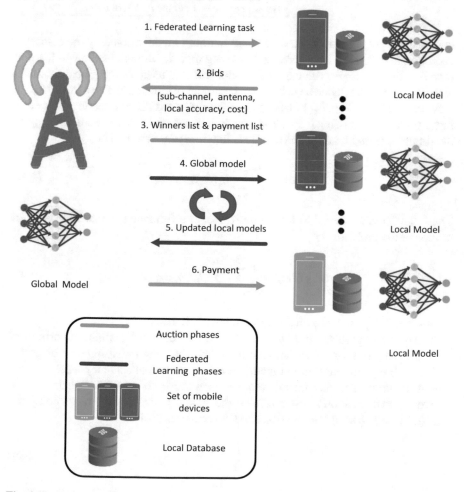

Fig. 4.19 System model

Table 4.4 Table of key
notations

Description	Notation
Global loss function	$L(\omega)$
Computing energy consumption of user n	E_n^{comp}
Computing time of one local iteration of user n	T_n^{comp}
Transmission time of user n	T_n^{com}
Transmission energy of user n	E_n^{com}
Total transmission time of user n	T_n^{tol}
Total energy consumption of user n	E_n^{tol}
Local accuracy of user n	ε_n
Computing resource of user n	f_n
Transmission power of user n	p_n
Antenna number desired by user n	A_n
The real cost of the ith bid of user n	V_{ni}
The claimed cost of the ith bid of user n	v_{ni}
The satisfaction level of the ith bid of user n	χ_{ni}
The winner indicator of the ith bid of user n	x_{ni}

is its corresponding output. The federated learning model trained by the dataset of each user is called the local federated learning model, while the federated learning model at the BS aggregates the local model from all users as the global federated learning model. We define a vector ω as the model parameter. We also introduce the loss function $l_n(\omega, a_{nk}, b_{nk})$ that captures the federated learning performance over input vector a_{nk} and output b_{nk}. The loss function may be different, depending on the different learning tasks. The total loss function of user n will be

$$L_n(\omega) = \frac{1}{s_n} \sum_{k=1}^{s_n} l_n(\omega, a_{nk}, b_{nk}). \tag{4.45}$$

Then, the learning model is the minimizer of the following global loss function minimization problem

$$\min_{\omega} \quad L(\omega) = \frac{1}{S} \sum_{n=1}^{N} \sum_{k=1}^{s_n} l_n(\omega, a_{nk}, b_{nk}), \tag{4.46}$$

where $S = \sum_{n=1}^{N} s_n$ is the total data samples of all users.

To solve the problem in (4.46), we adopt the federated learning algorithm of [72]. The algorithm uses an iterative approach that requires a number of global iterations (i.e., communication rounds) to achieve a global accuracy level. In each global iteration, there are interactions between the users and BS. Specifically, at a given global iteration t, users receive the global parameter ω^t, users computes $\nabla L_n(\omega^t)$, $\forall n$ and send it to the BS. The BS computes [70]

$$\nabla L(\omega^t) = \frac{1}{N} \sum_{n=1}^{N} \nabla L_n(\omega^t), \tag{4.47}$$

and then broadcasts the value of $\nabla L(\omega^t)$ to all participating users. Each participating user n will use local training data s_n to solve the local federated learning problem is defined as

$$\min_{\phi_n} \quad \mathcal{G}_n(\omega^t, \phi_n)$$

$$= L_n(\omega^t + \phi_n) - \left(\nabla L_n(\omega^t) - \varpi \nabla L(\omega^t)\right)^T \phi_n, \tag{4.48}$$

where ϕ_n represents the difference between global federated learning parameter and local federated learning parameter for user n. Each participating user n uses the gradient method to solve (4.48) with local accuracy ε_n that characterizes the quality of the local solution, and produces the output ϕ_n that satisfies

$$\mathcal{G}_n(\omega^t, \phi_n) - \mathcal{G}_n(\omega^t, \phi_n^*) < \varepsilon_n(\mathcal{G}_n(\omega^t, 0) - \mathcal{G}_n(\omega^t, \phi_n^*)). \tag{4.49}$$

Solving (4.48) also takes multiple local iterations to achieve a particular local accuracy. Then each user n sends the local parameter ϕ_n to the BS. Next, the BS aggregates the local parameters from the users and computes

$$\omega^{t+1} = \omega^t + \frac{1}{N} \sum_{n=1}^{N} \phi_n^t, \tag{4.50}$$

and broadcasts the value to all users, which is used for next iteration $t + 1$. This process is repeated until the global accuracy γ of (4.46) is obtained.

Assume that $L_n(\omega)$ is H-Lipschitz continuous and π-strongly convex, i.e.,

$$\pi I \preceq \nabla^2 L_n(\omega) \preceq H I, \forall n \in \mathcal{N},$$

the general lower bound on the number of global iterations is depends on local accuracy ε and the global accuracy γ as [70]:

$$I^g(\gamma, \varepsilon) = \frac{C_1 \log(1/\gamma)}{1 - \varepsilon}, \tag{4.51}$$

where the local accuracy measures the quality of the local solution as described in the preceding paragraphs.

In (4.51), we observe that a very high local accuracy (small ε) can significantly boost the global accuracy γ for a fixed number of global iterations I^g at the BS to solve the global problem. However, each user n has to spend excessive resources in terms of local iterations, I_n^l to attain a small value of ε_n. The lower bound on the number of local iterations needed to achieve local accuracy ε_n is derived as [70]

$$I_n^l(\varepsilon_n) = \vartheta_n \log \left(\frac{1}{\varepsilon_n}\right), \tag{4.52}$$

where $\vartheta_n > 0$ is a parameter choice of user n that depends on parameters of $L_n(\omega)$ [70]. In this section, we normalize $\vartheta_n = 1$. Therefore, to address this trade-off, the BS can set up an economic interaction environment to motivate the participating users to enhance local accuracy ε_n. Correspondingly, with the increased payment, the participating users are motivated to attain better local accuracy ε_n (i.e., smaller values), which as noted in (4.51) can improve the global accuracy γ for a fixed number of iterations I^g of the BS to solve the global problem. In this case, the corresponding performance bound in (4.51) for the heterogeneous responses ϵ_n can be updated to catch the statistical and system-level heterogeneity regarding the worst case of the participating users' responses as:

$$I^g(\gamma, \varepsilon_n) = \frac{\varpi \log(1/\gamma)}{1 - \max_n \varepsilon_n}, \forall n. \tag{4.53}$$

Computation and Communication Models for Federated Learning

The contributed computation resource that user n contributes for local model training is denoted as f_n. Then, c_n denotes the number of CPU cycles needed for the user n to perform one sample of data in local training. Thus, energy consumption of the user for one local iteration is presented as

$$E_n^{com}(f_n) = \zeta c_n s_n f_n^2, \tag{4.54}$$

where ζ is the effective capacitance parameter of computing chipset for user n. The computing time of a local iteration at the user n is denoted by

$$T_n^{comp} = \frac{c_n s_n}{f_n}. \tag{4.55}$$

It is noted that the uplink from the users to the BS is used to transmit the parameters of the local federated learning model while the downlink is used for transmitting the parameters of the global federated learning model. In this section, we just consider the uplink bandwidth allocation due to the relation of the uplink bandwidth and the cost that user experiences during learning a global model. We consider the uplink transmission of an OFDMA-based cellular system. A set of $\mathcal{B} = \{1, 2, ..., B\}$ subchannels each with bandwidth W. Moreover, the BS is equipped with A antennas and each user equipment has a single antenna (i.e., multi-user MIMO). We assume A to be large (e.g., several hundreds) to achieve massive MIMO effect which scales up traditional MIMO by orders of magnitude. Massive MIMO uses spatial-division multiplexing. In this section, we assume that the BS has perfect channel state information (CSI) and the channel gain is perfectly estimated, similar to [132, 133].

Then, the achievable uplink data rate of mobile user n is expressed as [133, 134]:

$$r_n = b_n W \log_2\left(1 + \frac{(A_n - 1)p_n h_n}{b_n W N_0}\right), \tag{4.56}$$

where p_n is the transmission power of user n, h_n is the channel gain of peer to peer link between user and the BS, N_0 is the background noise, A_n is the number of antennas the BS assigns to user n, and b_n is the number of sub-channels that user n uses to transmit the local model update to the BS.

We denote σ as the data size of a local model update and it is the same for all users. Therefore, the transmission time of a local model update is

$$T_n^{com}(p_n, A_n, b_n) = \frac{\sigma}{r_n}. \tag{4.57}$$

To transmit local model updates in a global iteration, the user n uses the amount of energy given as

$$E^{com}(p_n, f_n, A_n, b_n) = T^{com} p_n = \frac{\sigma p_n}{r_n}. \tag{4.58}$$

Hence, the total time of one global iteration for user n is denoted as

$$\begin{aligned} &T_n^{tol}(p_n, f_n, A_n, b_n, \varepsilon_n) \\ &= \log\left(\frac{1}{\varepsilon_n}\right) T_n^{comp}(f_n) + T_n^{com}(p_n, A_n, b_n). \end{aligned} \tag{4.59}$$

Therefore, the total energy consumption of a user n in one global iteration is denoted as follows

$$\begin{aligned} &E_n^{tol}(p_n, f_n, A_n, b_n, \varepsilon_n) \\ &= \log\left(\frac{1}{\varepsilon_n}\right) E_n^{comp}(f_n) + E_n^{com}(p_n, A_n, b_n). \end{aligned} \tag{4.60}$$

Auction Model

As described in Fig. 4.19, the BS first initializes the global network model. Then, the BS announces the auction rule and advertises the federated learning task to the mobile users. The mobile users then report their bids. Here, mobile user n submits a set of I_n of bids to the BS. A bid Δ_{ni} denotes the ith bid submitted by the mobile user n. Bid b_{ni} consists of the resource (sub-channel number b_{ni}, antenna number A_{ni}, local accuracy level ε_{ni}) and the claimed cost v_{ni} for the model training. Each mobile user n has its own discretion to determine its true cost V_{ni}, which will be presented in Sect. 4.3.1. Let x_{ni} be a binary variable indicating the bid Δ_{ni} wins or not. After receiving all the bids from mobile users, the BS decides winners and then allocates the resource to the winning mobile users. The winning mobile users join the federated learning and receive the payment after finishing the training model.

Remark In each bid, the bidder declares the requested resources, the local accuracy, and the corresponding cost. And the cost is calculated before submitting bids. Therefore, the cost corresponding to the requesting resources can be included in the bid during the bidding process.

Following we discuss one practical usage of our proposed auction scheme in federated learning. Let's consider a concrete example of a mobile phone keyboard such as Gboard (Google Keyboard). A large amount of local data will be generated when users interact with the keyboard app on their mobile devices. Suppose that Google server wants to train a next-word prediction model based on users' data. The server can announce the learning project to users through the app and encourage their participation. If a user wants to know more about this project, the app will display an interface to submit the bids and calculate the expected cost. If the user is interested in learning, he/she will download apps, calculate cost and submit the bids through the interface. Once the BS receive all bids in certain time, the BS will start the training process by broadcasting an initial global model to all the winning users. On behalf of the user, the app will download this global model and upload the model updates generated by the training on the user's local data. After finishing the model training project, the BS will give users rewards (e.g., money) based on the bid it wins.

Deciding Mobile Users's Bid

To transmit the local model update to the BS, mobile users need sub-channels and antenna resources. However, given the maximum tolerable time of federated learning, there is a correlation between resource and corresponding energy cost. In this section, we present the way mobile users decide bids. Specially, for bid Δ_{ni}, mobile user n calculates transmission power p_{ni}, computation resource f_{ni} and cost v_{ni} corresponding to a given sub-channel number b_{ni} and antenna number A_{ni}. However, for simplicity, the process to decide mobile users' bid is the same for every submitted bid. Thus, we remove the bid index i in this section. The energy cost of mobile user n is defined after user n solve the weighted sum of completion time and total energy consumption in the submitted bid, which is given as

$$\textbf{P1}: \quad \min_{f_n, p_n, A_n, b_n, \varepsilon_n} \quad I_0^n \left(E_n^{tol}(p_n, f_n, \varepsilon_n) + \rho T_n^{tol}(p_n, f_n, \varepsilon_n) \right) \tag{4.61a}$$

$$s.t. \quad I_0^n T_n^{tol}(p_n, f_n, A_n, b_n, \varepsilon_n) \leq T_{max}, \tag{4.61b}$$

$$f_n \in [f_n^{\min}, f_n^{\max}], \tag{4.61c}$$

$$p_n \in (0, p_n^{\max}], \tag{4.61d}$$

$$\varepsilon_n \leq (0, 1], \tag{4.61e}$$

$$A_n \in (0, A_n^{\max}], \tag{4.61f}$$

$$b_n \in (0, b_n^{\max}], \tag{4.61g}$$

where $f_n^{\max}$ and $p_n^{\max}$ are the maximum local computation capacity and maximum transmit power of mobile user n, respectively. $A_n^{\max}$ and $b_n^{\max}$ are the maximum antenna and maximum sub-channel that mobile user n can request in each bid, respectively. $A_n^{\max}$ and $b_n^{\max}$ are chosen by mobile user n. $I_0^n = \frac{C_1 \log(1/\gamma)}{1-\varepsilon_n}$ is the lower bound of the number global iterations corresponding to local accuracy ε_n. Note that the cost to the mobile user cannot be the same over iterations. ρ is the weight. However, to make the problem more tractable, we consider minimizing the approximated cost rather than the actual cost, similar to approach in [111]. Constraint (4.61b) indicates delay requirement of federated learning task.

According to **P1**, the maximum number of antennas and sub-channels are always energy efficient, i.e., the optimal antenna is $A_n = A_n^{max}$, $b_n = b_n^{max}$ and $\varepsilon_n^*, p_n^*, f_n^*$ are the optimal solution to:

$$\mathbf{P2}: \quad \min_{f_n, p_n, \varepsilon_n} \quad I_0^n \left(E_n^{tol}(p_n, f_n, \varepsilon_n) + \rho T_n^{tol}(p_n, f_n, \varepsilon_n) \right)$$

$$\text{s.t.} \quad I_0^n T_n^{tol}(p_n, f_n, \varepsilon_n) \leq T_{max},$$

$$f_n \in [f_n^{\min}, f_n^{\max}], \tag{4.62}$$

$$\varepsilon_n \in (0, 1],$$

$$p_n \in (0, p_n^{\max}].$$

Because of the non-convexity of **P2**, it is challenging to obtain the global optimal solution. To overcome the challenge, an iterative algorithm with low complexity is proposed in the following subsection.

Iterative Algorithm

The proposed iterative algorithm basically involves two steps in each iteration. To obtain the optimal, we first solve (**P2**) with fixed ε_n, and then ε_n is updated based on the obtained f_n, p_n in the previous step. In the first step, we consider the first case

---

Algorithm 6 Optimal uplink power transmission

1: Calculate $\phi(p_n^{max})$
2: Calculate p_n^{min} so that $I_0^n T_n^{tol}(p_n^{min}) = T_{max}$
3: **if** $\phi(p_n^{max} < 0)$ **then**
4:    $p_n^* = p_n^{max}$
5: **else**
6:    $p_1 = \max(0, p_n^{min})$ and $p_2 = p_n^{max}$
7:    **while** $(p_2 - p_1 \leq \epsilon)$ **do**
8:       $p_u = (p_1 + p_1)/2$
9:       **if** $\phi(p_u) \leq 0$ **then**
10:          $p_1 = p_u$
11:      **else**
12:          $p_2 = p_u$
13:      **end if**
14:   **end while**
15:   $p_n^* = (p_1 + p_2)/2$
16: **end if**

---

when ε_n is fixed, and **P2** becomes

$$\textbf{P3}: \quad \min_{f_n, p_n, \varepsilon_n} \quad I_0^n \left(E_n^{tol}(p_n, f_n, \varepsilon_n) + \rho T_n^{tol}(p_n, f_n, \varepsilon_n) \right)$$

$$s.t. \quad I_0^n T_n^{tol}(p_n, f_n, \varepsilon_n) \leq T_{max}, \tag{4.63}$$

$$f_n \in [f_n^{min}, f_n^{max}],$$

$$p_n \in (0, p_n^{max}].$$

P3 can be decomposed into two sub-problems as follows.

Optimization of Uplink Transmission Power

Each mobile user assigns its transmission power by solving the following problem:

$$\textbf{P3a}: \quad \min_{p_n} \quad f(p_n)$$

$$s.t. \quad I_0^n T_n^{tol}(p_n, f_n, \varepsilon_n) \leq T_{max}, \tag{4.64}$$

$$p_n \in (0, p_n^{max}],$$

$$f_n, \varepsilon_n \text{ are given.}$$

where $f(p_n) = \frac{\sigma(1+\rho)p_n}{b_n W \log_2(1+\frac{(A_n-1)p_n h_n}{b_n W N_0})}$.

Algorithm 7 Optimal local accuracy

1: Initialize $\varepsilon_n = \varepsilon_n^{(0)}$, set $j = 0$
2: **repeat**
3: Calculate $\varepsilon_n^* = \frac{\alpha_1}{(\ln 2)\xi^j}$
4: Update $\xi^{(j+1)} = \frac{\gamma_1 \log_2(1/\varepsilon_n) + \gamma_2}{1 - \varepsilon_n}$
5: Set $j = j + 1$
6: **until** $|H(\xi^{(n+1)})|/|H(\xi^{(n)})| < \epsilon_2$

Note that $f(p_n)$ is quasiconvex in the domain [135]. A general approach to the quasiconvex optimization problem is the bisection method, which solves a convex feasibility problem each time [116]. However, solving convex feasibility problems by an interior cutting-plane method requires $O(\kappa^2/\alpha^2)$ iterations, where κ is the dimension of the problem [135]. On the other hand, we have

$$f'(p_n) = \frac{\sigma \log_2(1 + \theta_n p_n h_n) + \frac{\sigma(1+\rho)p_n\theta_n h_n}{\ln 2(1+\theta_n p_n h_n)}}{b_n W (\log(1 + \theta_n p_n h_n))^2}, \tag{4.65}$$

where $\theta_n = \frac{(A_n - 1)}{W N_0}$. Then, we have

$$\phi(p_n) = \sigma \log_2(1 + \theta_n p_n h_n) + \frac{\sigma(1 + \rho)p_n\theta_n h_n}{\ln 2(1 + \theta_n p_n h_n)} \tag{4.66}$$

is a monotonically increasing transcendental function and negative at the starting point $p_n = 0$ [135]. Therefore, in order to obtain the optimal power allocation p_n as shown in Algorithm 6, we follow a low-complexity bisection method by calculating $\phi(p_n)$ rather than solving a convex feasibility problem each time.

Optimization of CPU Cycle Frequency and Number of Antennas

$$\textbf{P3b}: \quad \min_{f_n} \quad I_0^n \log\left(\frac{1}{\varepsilon_n}\right) c_n s_n \left(\varsigma f_n^2 + \rho/f_n\right)$$

$$\text{s.t.} \quad I_0^n \left(\log\left(\frac{1}{\varepsilon_n}\right) \frac{c_n s_n}{f_n} + T_n^{com}\right) \leq T_{max}, \tag{4.67}$$

$$f_n \in [f_n^{min}, f_n^{max}],$$

$$p_n, \varepsilon_n \text{ are given.}$$

P3b is the convex problem, so we can solve it by any convex optimization tool.

Algorithm 8 Iterative algorithm

1: Initialize a feasible solution p_n, f_n, ε_n and set $j = 0$.
2: **repeat**
3:      With $\varepsilon_n^{(j)}$ obtain the optimal $p_n^{(j+1)}$, $f_n^{(j+1)}$ of problem **P2** :
4:      With $p_n^{(j+1)}$, $f_n^{(j+1)}$ obtain the optimal $\varepsilon_n^{(j+1)}$ of problem **P2** :
5:      Set $j = j + 1$
6: **until** Objective value of **P2** converges

In the second step, **P2** can be simplified by using f_n and p_n calculated in the first step as:

$$\textbf{P4}: \quad \min_{\varepsilon_n} \quad \frac{\gamma_1 \log_2(1/\varepsilon_n) + \gamma_2}{1 - \varepsilon_n} \tag{4.68a}$$

$$s.t. \quad T_n^{tol} \leq T_{max}, \tag{4.68b}$$

where $\gamma_1 = a\left(E_n^{comp} + T_n^{comp}\right)$ and $\gamma_2 = a\left(E_n^{com} + T_n^{com}\right)$. The constraint (4.68b) is equivalent to $T_n^{com} \leq \vartheta(\varepsilon_n)$, where $\vartheta(\varepsilon_n) = \frac{1 - \varepsilon_n}{m} T_{max} + \frac{c_n s_n \log_2 \varepsilon_n}{f_n}$. We have $\vartheta(\varepsilon_n)'' < 0$, and therefore, $\vartheta(\varepsilon_n)$ is a concave function. Thus, constraint (4.68b) can be equivalent transformed to $\varepsilon_n^{min} \leq \varepsilon_n \leq \varepsilon_n^{max}$, where $\vartheta(\varepsilon_n^{min}) = \vartheta(\varepsilon_n^{max}) = T_n^{com}$. Therefore, ε_n is the optimal solution to

$$\textbf{P5}: \quad \min_{\varepsilon_n} \quad \frac{\gamma_1 \log_2(1/\varepsilon_n) + \gamma_2}{1 - \varepsilon_n} \tag{4.69}$$

$$s.t. \quad \varepsilon_n^{min} \leq \varepsilon_n \leq \varepsilon_n^{max}.$$

Obviously, the objective function of **P5** has a fractional in nature, which is generally difficult to solve. According to [70, 136], solving **P5** is equivalent to finding the root of the nonlinear function $H(\xi)$ defined as follows

$$H(\xi) = \min_{\varepsilon_n^{min} \leq \varepsilon_n \leq \varepsilon_n^{max}} \gamma_1 \log_2(1/\varepsilon_n) + \gamma_2 - \xi(1 - \varepsilon_n) \tag{4.70}$$

Function $H(\xi)$ with fixed ξ is convex. Therefore, the optimal solution ε_n can be obtained by setting the first-order derivative of $H(\xi)$ to zero, which leads to the optimal solution is $\varepsilon_n^* = \frac{\gamma_1}{(\ln 2\xi)}$. Thus, similar to [70], problem **P5** can be solved by using the Dinkelbach method in [136] (shown as Algorithm 7).

Convergence Analysis

The algorithm that solves problems **P2** is given in Algorithm 4, which iteratively solves problems **P3** and **P4**. Since the optimal solution of problem **P3** and **P4** is obtained in each step, the objective value of problem **P2** is non-increasing in each step. Moreover, the objective value of problem **P2** is lower bounded by zero. Thus, Algorithm 4 always converges to a local optimal solution.

Complexity Analysis

Because of the non-convexity of **P2**, it is challenging to obtain the global optimal solution. To overcome the challenge, an iterative algorithm with low-complexity is proposed in the following subsection. In particular, to solve the general energy-efficient resource allocation problem **P2** using Algorithm 3, the major complexity in each step lies in solving problems **P3** and **P4**. To solve problem **P3**, the complexity is $\mathcal{O}(L_e \log_2(1/\epsilon_1))$, where ϵ_1 is the accuracy of solving **P3** with the bisection method and L_e is the number of iterations for optimizing f_n and p_n. To solve problem **P4**, the complexity is $\mathcal{O}(\log_2(1/\epsilon_2))$ with accuracy ϵ_2 by using the Dinkelbach method [136]. As a result, the total complexity of the proposed Algorithm 4 is $H_e S$, where H_e is the number of iterations for problems **P3** and **P4** and S is equal to $\mathcal{O}(L_e \log_2(1/\epsilon_1)) + \mathcal{O}(\log_2(1/\epsilon_2))$.

After deciding the bids, the mobile users submit bids to the BS. The following section describes the auction mechanism between the BS and mobile users for selecting winners, allocating bandwidth and deciding on payment.

4.3.2 Auction Mechanism Between BS and Mobile Users

After receiving all bids submitted by mobile users, the BS decides a set of winners by solving the problem (**P6**), aiming to maximize social welfare. The BS's aim is to achieve social welfare because the BS needs to incentive mobile users to participate in learning. Here, the BS's freedom in designing the incentive mechanism is the payment determination, which can force participant mobile users to be truthful. Moreover, if the BS wants to select winners to maximize its utility, the BS needs to know the distribution of mobile users' private information in advance[129], which is assumed to be unavailable in our work. In case the prior distribution of mobile users' private information is not available, worst-case analysis can be applied, but that method could lead to overly pessimistic results [129].

Problem Formulation

In bid Δ_{ni} that mobile user n submits to the BS includes the number of subchannels b_{ni}, the number of antennas A_{ni}, local accuracy ϵ_{ni}, and claimed cost v_{ni}. The utility of one bid is the difference between the payment g_{ni} and the real cost V_{ni}.

$$U_{ni} = \begin{cases} g_{ni} - V_{ni}, & \text{if bid } \Delta_{ni} \text{ wins,} \\ 0, & \text{otherwise.} \end{cases} \tag{4.71}$$

The payment that the BS pays for winning bids is $\sum_{n,i} g_{ni}$. As we described in Sect. 4.3.1, high local accuracy will significantly improve the global accuracy for a fixed number of global iterations. The utility of the BS is the difference between the BS's satisfaction level and the payment for mobile users. The satisfaction level of the BS to bid Δ_{ni} is measured based on the local accuracy that mobile user n can provide in the ith bid and is defined as follows

$$\chi_{ni} = \frac{\tau}{\varepsilon_{ni}}. \tag{4.72}$$

Thus, the total utilities of the system or the social welfare is

$$\sum_{n,i} (\chi_{ni} - v_{ni}) x_{ni}. \tag{4.73}$$

If mobile users truthfully submit their cost, $V_{ni} = v_{ni}$, we have the social welfare maximization problem defined as follows:

$$\mathbf{P6}: \quad \max_{x} \quad \sum_{n,i} (\chi_{ni} - v_{ni}) x_{ni} \tag{4.74a}$$

$$s.t. \quad \sum_{n} x_{ni} b_{ni} \leq B_{max}, \tag{4.74b}$$

$$\sum_{n} x_{ni} A_{ni} \leq A_{max}, \tag{4.74c}$$

$$\sum_{i} x_{ni} \leq 1, \forall n, \tag{4.74d}$$

$$x_{ni} = \{0, 1\}, \tag{4.74e}$$

where (4.74b) and (4.74c) indicate the bandwidth resource (i.e., sub-channels) and the antennas limitation constraints of the BS, respectively. Then, (4.74d) shows that a mobile user can win at most one bid and (4.74e) is the binary constraint that presents whether bid Δ_{ni} wins or not.

Problem **P6** is a minimization knapsack problem, which is known to be NP-hard. This implies that no algorithm is able to find out the optimal solution of **P6** in polynomial time. It is also known that a mechanism with Vickrey-Clarke-Groves (VCG) payment rule is truthful only when the resource allocation is optimal. Hence, using VCG payment directly is unsuitable due to the problem **P6** is computationally intractable. To deal with the NP-hard problem, we proposed the primal-dual based greedy algorithm. The following economic properties are desired.

Truthfulness An auction mechanism is truthful if and only if for every bidder n can get the highest utility when it reports true value.

Individual Rational If each mobile user reports its true information (i.e., cost and local accuracy), the utility for each bid is nonnegative, i.e., $U_{ni} \geq 0$.

Computation Efficiency The problem can be solved in polynomial time.

Among these three properties, truthfulness is the most challenging one to achieve. In order to design a truthful auction mechanism, we introduce the following definitions.

Definition 1 (Monotonicity) If mobile user n wins with the bid $\Delta_{ni} = \{v_{ni}, 1/\varepsilon_{ni}, b_{ni}, A_{ni}\}$, then mobile user n can win the bid with $\Delta_{nj} = \{v_{nj}, 1/\varepsilon_{nj}, b_{nj}, A_{nj}\} \succ \Delta_{ni} = \{v_{ni}, 1/\varepsilon_{ni}, b_{ni}, A_{ni}\}$.

The notation $\succ$ denotes the preference over bid pairs. Specifically, $\Delta_{nj} = \{v_{nj}, 1/\varepsilon_{nj}, b_{nj}, A_{nj}\} \succ \Delta_{ni} = \{v_{ni}, 1/\varepsilon_{ni}, b_{ni}, A_{ni}\}$ if $\varepsilon_{nj} < \varepsilon_{ni}$ for $v_{nj} = v_{ni}, b_{nj} = b_{ni}, A_{nj} = A_{ni}$ or $v_{nj} < v_{ni}, b_{nj} < b_{ni}, A_{nj} < A_{ni}$ for $\varepsilon_{nj} = \varepsilon_{ni}$. The monotonicity implies that the chance to obtain a required bundle of resources can only be enhanced by either increasing the local accuracy or decreasing the amount of resources required or decreasing the cost.

Definition 2 (Critical Value) For a given monotone allocation scheme, there exists a critical value c_{ni} of each bid Δ_{ni} such that $\forall n, i(\chi_{ni} - v_{ni}) \geq c_{ni}$ will be a winning bid, while $\forall n, i(\chi_{ni} - v_{ni}) < c_{ni}$ is a losing bid.

In our proposed mechanism, the difference between the satisfaction based on local accuracy and cost of one bid can be considered as the value of that bid. Therefore, the critical value can be seen as the minimum value that one bidder has to bid to obtain the requested bundle of resources. With the concepts of monotonicity and critical value, we have the following lemma.

Lemma 4.1 *An auction mechanism is truthful if the allocation scheme is monotone and each winning mobile user is paid the amount that equals to the difference between the satisfaction based on the local accuracy and the critical value.*

Proof Similar Lemma 1 and Theorem 1 in [130].

In the next subsection, we propose a primal-dual greedy approximation algorithm for solving problem **P6**. The algorithm iteratively updates both primal and dual variables and the approximation analysis is based on duality property. As the result, we firstly relax $1 \geq x_{ni} \geq 0$ of **P6** to have the linear programming relaxation (LPR) of **P6**. Then, we introduce the dual variable vectors y, z and t corresponding to constraints (4.74b), (4.74c) and (4.74d) and we have the dual of problem LPR of **P6** can be written as

$$\textbf{P7}: \quad \max_{\textbf{y,z,t}} \quad \sum_{n \in \mathcal{N}} y_n + zB_{max} + tA_{max} \tag{4.75a}$$

$$s.t. \quad y_n + zA_{ni} + tB_{ni} \geq q_{ni}, \forall n, i, \tag{4.75b}$$

$$y_n \geq 0, \forall n, \tag{4.75c}$$

$$z, t \geq 0. \tag{4.75d}$$

In Sect. 4.3.2, we devise an greedy approximation algorithm and Sect. 4.3.2, a theoretical bound is achieved for the approximation ratio of the proposed algorithm.

Approximation Algorithm Design

In this section, we use a greedy algorithm to solve problem **P6** The main idea of the greedy algorithm is to allocate the resource to bidders with the larger normalized value. The winner selection process is described in the Algorithm 4. The process consists 3 steps:

Step 1:   Based on the bid's value and the weighted sum of requested resources, each bid Δ_{ni} calculates the normalized value. The bid's value is defined as the difference between the satisfaction level of the BS and the cost declared in this bid, $q_{ni} = \chi_{ni} - v_{ni}$. The weighted sum of different types of resources declared in this bid is defined as $s_n^i = \eta_b B_n + \eta_a A_n$, where η_b, η_a are the weights. The normalized value of the bid is defined as the ration between the value of this bid and the weighted sum of requested resources, and is denoted as

$$\bar{q}_{ni} = \frac{q_{ni}}{s_{ni}}.$$

Step 2:   The bid with maximum $\bar{q}_{ni}$ wins the bidding.
Step 3:   Delete user n from the list of bidders. Then go back to Step 2 until either one of the following termination conditions is satisfied:

  (i) The BS has not enough resource to satisfy the demand;
  (ii) All the mobile users win one bid.

Approximation Ratio Analysis

In this subsection, we analyze approximation ratio of Algorithm 10. Our approach is to use the duality property to derive a bound for approximation algorithm. We denote the optimal solution and the optimal value of LPR of **P6** as x_{ni}^* and OP_f. Furthermore, let OP and φ as the optimal value of **P6** and the primal value of **P6** obtained by Algorithm 10. Our analysis consists of two steps. First, Theorem 4.1 shows that Algorithm 10 generates a feasible solution to **P7**, and Proposition 1 provides approximation factor.

Theorem 4.1 *Algorithm 10 provides a feasible solution to **P7**.*

Proof We discuss the following three cases:

- Case 1: mobile user μ wins, i.e., $\mu \in \mathcal{U}$ and $b_{\mu i_\mu} = \max_{i' \in \mathcal{I}_\mu}\{q_{\mu i'}\}$. Then we have $y_\mu = q_{\mu i_\mu} \geq q_{\mu i'}, \forall i' \in \mathcal{I}_\mu$. Thus, constraint (4.75b) is satisfied for all mobile users in $\mathcal{U}$.

Algorithm 9 The Greedy approximation algorithm

1: **Input:** $(B, A, \chi, v, B_{max}, A_{max})$
2: **Output:** solution $\mathbf{x}$
3: $\mathcal{U} = \varnothing, \mathbf{x} = \mathbf{0}$
4: $\forall n : y_n = 0, \psi = 0;$
5: $\varphi = 0, B = 0, A = 0;$
6: $s_n^i = \eta_b B_{ni} + \eta_a A_{ni};$
7: $q_{kj} = \chi_{ni} - v_{ni};$
8: **for** $n \in \mathcal{N}$ **do**
9: $\quad i_n = \arg \max_i \{q_{ni}\};$
10: **end for**
11: $\kappa = \max \frac{s_{ni}}{s_{ni'}};$
12: **while** $\mathcal{N} \neq \varnothing$ **do**
13: $\quad \mu = \arg \max_{n \in \mathcal{N}} \frac{q_{ni}}{s_{nin}};$
14: $\quad$ **if** $B + b_{\mu i_\mu} <= B_{max}$ and $A + a_{\mu i_\mu} <= A_{max}$ **then**
15: $\quad\quad x_{\mu i_\mu} = 1; \quad y_\mu = q_{\mu i_\mu};$
16: $\quad\quad \varphi = \varphi + q_{\mu i_\mu};$
17: $\quad\quad \psi = \frac{\sum_{n \in \mathcal{U}} q_{nin}}{\sum_{n \in \mathcal{U}} s_{nin}};$
18: $\quad\quad \mathcal{U} = \mathcal{U} \cup \{\mu\}$ and $\mathcal{N} = \mathcal{N} \setminus \{\mu\}$
19: $\quad$ **else**
20: $\quad\quad$ break;
21: $\quad$ **end if**
22: **end while**
23: $\bar{\psi} = \kappa \psi;$
24: $z = \eta_b \bar{\psi}, \quad t = \eta_a \bar{\psi}$

- Case 2: mobile user μ loses the auction, i.e., $\mu \in \mathcal{N} \setminus \mathcal{U}$. According to the while loop, it is evident that

$$\frac{q_{nin}}{s_{nin}} > \frac{q_{\mu i_\mu}}{s_{\mu i_\mu}}, \forall n \in \mathcal{U}.$$

Therefore, $\psi > \frac{q_{\mu i_\mu}}{s_{\mu i_\mu}}$. Thus,

$$\bar{\psi} \geq \kappa \frac{q_{\mu i_\mu}}{s_{\mu i_\mu}} \geq \frac{q_{\mu i_\mu}}{s_{\mu i_\mu}}.$$

In addition, we have

$$q_{\mu i_\mu} \geq q_{\mu i'} \quad \text{and} \quad \kappa > \frac{s_{\mu i_\mu}}{s_{\mu i'}}, \forall i' \neq i_\mu.$$

Therefore,

$$\bar{\psi} \geq \frac{q_{\mu i'}}{n_{\mu i'}}, \forall i' \neq i_\mu.$$

Therefore, we have

$$\eta_b \bar{\psi} B_{in} + \eta_a \bar{\psi} A_{in} \geq q_{in}, \forall i' \neq i_\mu.$$

or

$$z C_{in} + t A_{in} \geq q_{in}, \forall i' \neq i_\mu.$$

Therefore, constraint (4.75b) is also satisfied for all mobile users in $\mathcal{N} \setminus \mathcal{U}$.

Proposition 1 *The upper bound of integrality gap α between* **P6** *and its relaxation and the approximation ratio of Algorithm 10 are* $1 + \frac{\kappa \Upsilon}{\Upsilon - S}$, *where* $\Upsilon = \eta_b B_{max} + \eta_a A_{max}$, $S = \max_{n,i} s_{ni}$.

Proof Let OP and OP_f be the optimal solution for **P6** and LPR of **P6**. We can obtain the following:

$$OP \leq OP_f \leq \sum_{n=1}^{N} y_n + z B_{max} + t A_{max}$$

$$\leq \sum_{n=1}^{N} y_n + \bar{\psi}(\eta_b B_{max} + \eta_a A_{max})$$

$$\leq \sum_{n \in \mathcal{N}} q_{ni_n} + \bar{\psi}(\eta_b B_{max} + \eta_a A_{max})$$

$$\leq \left(\sum_{n \in \mathcal{N}} q_{ni_n}\right) \left(1 + \frac{(\eta_b B_{max} + \eta_a A_{max})\kappa}{\eta_b B_{max} + \eta_a A_{max} - S}\right)$$

$$\leq \varphi \left(1 + \frac{\Upsilon \kappa}{\Upsilon - S}\right),$$

Therefore, the integrality α is given as

$$OP_f / OP$$
$$\leq \quad OP_f / \varphi$$
$$\leq \quad \left(1 + \frac{\kappa \Upsilon}{\Upsilon - S}\right).$$

The approximation ratio is

$$OP/\varphi \leq OP_f/\varphi \leq \left(1 + \frac{\kappa \Upsilon}{\Upsilon - S}\right).$$

Payment

Then we will find the critical value which is the minimum value a bidder has to bid to win the requested bundle of resources. In this section, we consider the bid combinations submitted by mobile user n as the combinations of bids submitted by virtual bidders, in which each virtual bidder can submit one bid. Therefore, the number of virtual bidders corresponding to mobile user n is equal to the number of bids I_n that mobile user n submits. Denote by m the losing mobile user with the highest normalized value if mobile user n is not participating in the auction. Accordingly, the minimum value mobile user n needs to place is $\frac{q_{mi_m}}{s_{mi_m}} s_{ni_n}$, where i_m and i_n are the indexes of highest normalized value bids of mobile user m and n, respectively. Thus, the payment of winning mobile user n in the pricing scheme is $g_{ni_n} = \chi_{ni_n} - \frac{q_{mi_m}}{s_{mi_m}} s_{ni_n}$.

Properties

Now, we show that the winner determination algorithm is monotone and the payment determined for a winner mobile user is the difference between the local accuracy-based satisfaction and the critical value of its bid. From line 13 of the Algorithm 10, it is clear that a mobile user can increase its chance of winning by increasing its bid. Also, a mobile user can increase its chance to win by decreasing the weighted sum of the resources. Therefore, the winner determination algorithm is monotone with respect to mobile user's bids. Moreover, the value of a winning bidder is equals to the minimum value it has to bid to win its bundle, i.e., its critical value. This is done by finding the losing bidder m who would win if bidder n would not participate in the auction. Thus, the proposed mechanism has a monotone allocation algorithm and payment for the winning bidder equals the difference between the local accuracy-based satisfaction and the critical value of its bid. We conclude that the proposed mechanism is a truthful mechanism according to Lemma 4.1.

Next, we prove that the proposed auction mechanism is individually rational. For any mobile user n bidding its true value, we consider two possible cases:

- If mobile user n is a winner with its bid ith, its payment is

$$U_{ni} = g_{ni} - v_{ni}$$

$$= (\chi_{ni} - \frac{q_{mi_m}}{s_{mi_m}} s_{ni} - v_{ni}$$

$$= \left(\frac{\chi_{ni} - v_{ni}}{s_{ni}} - \frac{q_{mi_m}}{s_{mi_m}} \right) s_{ni}$$

$$= \left(\frac{q_{ni}}{s_{ni}} - \frac{q_{mi_m}}{s_{mi_m}} \right) s_{ni} \geq 0$$

where m the losing bidder with the highest normalized valuation if n does not participate in the auction and the last inequality follows from Algorithm 10.
- If mobile user n is not a winner. Its utility is 0.

Therefore, the proposed auction mechanism is individually rational.

Finally, we show that the proposed auction mechanism is computationally efficient. We can see that in Algorithm 10, the while-loop (lines 12–22) takes at most N times, linear to input. Calculating the payment takes at most $N(N-1)$ times. Therefore, the proposed auction mechanism is computationally efficient. Therefore, the time complexity of Algorithm 10 is $\mathcal{O}(N^2)$.

4.3.3   Simulations

In this section, we provide some simulation results to evaluate the proposed mechanism. The parameters for the simulation are set the following. The required CPU cycles for performing a data sample c_n is uniformly distributed between $[10, 50]$ cycles/bit [70]. The size of data samples of each mobile user is $s_n = 80 \times 10^6$. The effective switched capacitance in local computation is $\xi = 10^{-28}$ [70]. We assume that the noise power spectral density level N_0 is -174dBm/Hz, the sub-channel bandwidth is $W = 15$ kHz and the channel gain is uniformly distributed between $[-90, -95]$ dB [132]. In addition, the maximum and minimum transmit power of each mobile user is uniformly distributed between $[6, 10]$ mW and between $[0, 2]$ mW, respectively. The maximum and minimum computation capacity is uniformly distributed between $[3, 5]$ GHz and between $[10, 20]$ Hz, respectively. We also assume that the total number of sub-channels and antennas of the BS are 100 and 100, respectively.

Firstly, we use the iterative Algorithm 8 to perform the characteristic of evaluating bids when $\rho = 1$. The maximum number of sub-channels B_n^{max} and antennas A_n^{max} for mobile user n to request in each bid vary from 10 to 50. Figure 4.20 shows the accuracy level that mobile user n requires to provide increases when the maximum number of sub-channels B_n^{max} and antennas A_n^{max} increase. In particular, when the sub-channels and antennas are both 50, the local accuracy 0.92 while when the sub-channels and antennas are both 10, the local accuracy 0.81. This is because the transmission time and transmission cost decrease when wireless resources increase. It requires less global round to satisfy the learning task performance. Therefore, the local accuracy increases or . As shown in Fig. 4.21, the energy cost decreases when the number of sub-channels and antennas increases. This is because mobile user n can keep low contributing CPU cycle frequency and transmission rate while guaranteeing the delay constraint.

Figures 4.22 and 4.23 present the cost of one bid of the mobile user and local accuracy, respectively, when the weight ρ varies from 1 to 9. As shown in Figs. 4.22 and 4.23, when ρ increases, the local accuracy decreases and the energy cost increases. This is because when ρ increases, the objective focuses more on

Fig. 4.20 Changes local accuracy when the maximum number sub-channels and antennas in one bid vary

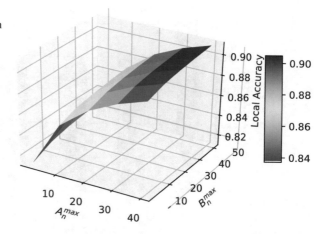

Fig. 4.21 Changes in energy cost when the maximum number sub-channels and antennas in one bid vary

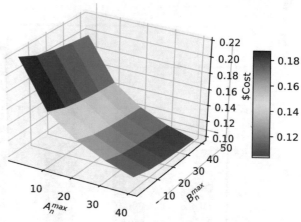

minimizing the time completion of one global round. It requires more computation resources as well as better quality of data (low local accuracy).

In the following, we evaluate the performance of the proposed auction algorithm. To compare with the proposed algorithm, we use four baselines:

- Optimal Solution: **P6** is solved optimally.
- Fixed Price Scheme [137]: In this scheme, price vector $f = \{f_b, f_a\}$ is the price mobile users need to pay for the resource. In this scheme, the mobile users are served in a first-come, first-served basic until the resources are exhausted. The mobile user can get the resource when the valuation of mobile user's bid is at least $F_{ni} = B_{ni} f_b + A_{ni} f_a$ which is the sum of the fixed price of each resource in their bid. We consider three kinds of price vector: linear price ($f_i = f_o \times \eta_i, i = a, b$), sub-linear price vector ($f_i = f_o \times \eta_i^{0.85}, i = a, b$) [137], and a super-linear price vector ($f_i = f_o \times \eta_i^{1.15}, i = a, b$) [137]. Here, we call f_o as the basic price. Unless specified otherwise, we choose $f_o = 0.01$.

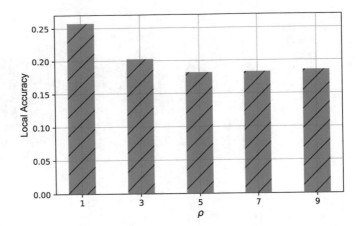

Fig. 4.22 Local accuracy v.s. ρ

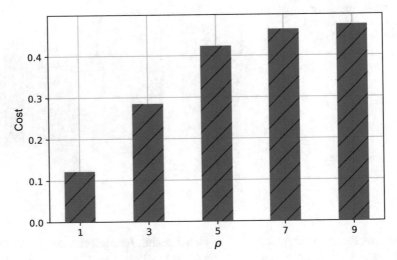

Fig. 4.23 Energy cost v.s. ρ

- Reward-based greedy auction [138].
- Maximum utility of the BS.

Figure 4.24 reports the performance of the optimal solution, the lower bound, and the proposed greedy scheme. The lower bound is determined by the fractional optimal solution divided by gap when the number of mobile users varies from 20 to 100 with a step size of 20. We note that with the number of mobile users increasing, all schemes produce higher social welfare. This is because there are more chances to choose winning bids with a higher value. Although the social welfare obtained through the proposed greedy scheme is lower than through optimal solution and much higher than the lower bound.

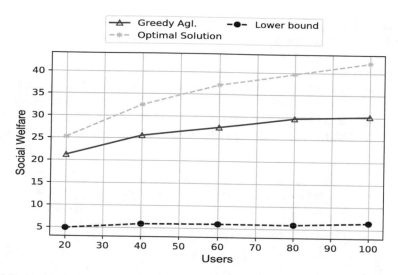

Fig. 4.24 Social welfare vs. users

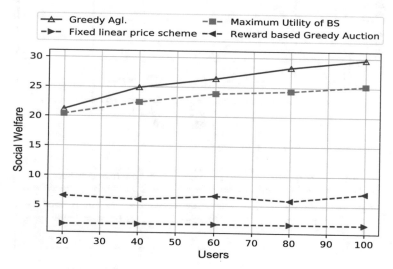

Fig. 4.25 Social welfare vs. users

Figure 4.25 shows the social cost achieved by the proposed greedy scheme, the maximum utility of the BS, reward-based greedy auction, and fixed linear scheme when the number of mobile users varies 20–100 with the step size of 20. We can see that the proposed greedy scheme can provide much higher social welfare than the baseline. When the number of users is 100, the social welfare obtained by a proposed greedy algorithm is approximately 16% higher than the one obtained by the maximum utility of BS. The result is that our proposed algorithm focuses on maximizing social welfare. In addition, when the number of users is 100, the social welfare obtained by the proposed greedy algorithm is approximately 4 times and 15 times higher than the one obtained by the reward-based Greedy Auction and fixed linear price scheme, which ignores the wireless resource limitation when deciding the winning bids.

Since the fixed price scheme heavily depends on the prices of resources, the next experiment helps us to decide whether the fixed-price vector or the performance of the proposed mechanisms is better when we change the basic price f_o between $[0.01, 0.31]$ with the step is 0.03. Figures 4.26, 4.27, and 4.28 show that the social

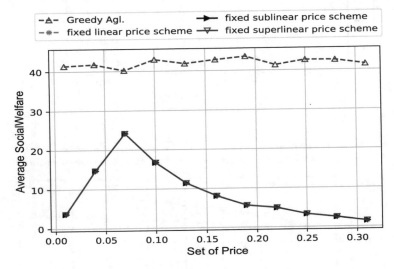

Fig. 4.26 Social welfare for $\eta_a = 1, \eta_b = 0.5$

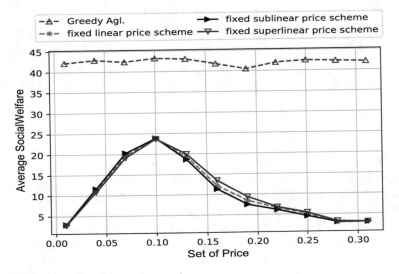

Fig. 4.27 Social welfare for $\eta_a = 1, \eta_b = 1$

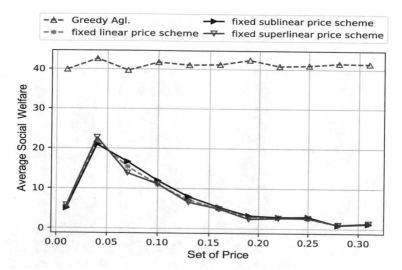

Fig. 4.28 Social welfare for $\eta_a = 1, \eta_b = 2$

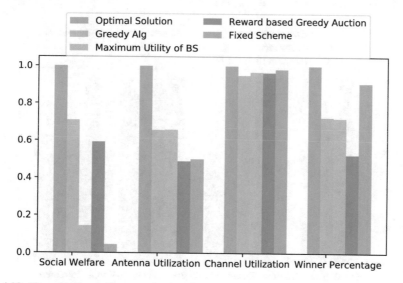

Fig. 4.29 Normalized ratio for $\eta_a = 1, \eta_b = 0.5$

welfare of fixed price firstly increases and then decreases and equal to 0 when the initial price increases. This is because when the basic price becomes too high, the sum of the price is higher than the valuation of the resources claimed in a bid. Moreover, the social welfare achieved by linear, sublinear, and superlinear price schemes is lower than by the proposed greedy scheme. This proves our proposed auction scheme outperforms the fixed price scheme.

In Figs. 4.29, 4.30, and 4.31, we observe the metrics: social welfare, resource utilization and percentage of five schemes: greedy proposed scheme and fixed price

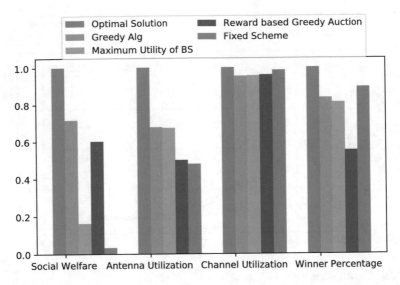

Fig. 4.30 Normalized ratio for $\eta_a = 1, \eta_b = 1$

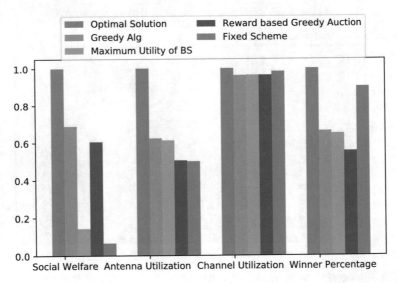

Fig. 4.31 Normalized ratio for $\eta_a = 1, \eta_b = 2$

schemes with other baselines. We perform in terms of the ratio with the proposed greedy scheme. Among these schemes, the optimal solution is the highest in terms of all metrics. Compared with the proposed scheme, the fixed price can utilize more resources and more mobile users but provides less social welfare. This is due to the fact that the fixed price mechanism heavily depends on the prices of the resources. In addition, the resource utilization of our proposed scheme is competitive to the one of the maximum utility of the BS scheme and the reward-based greedy auction scheme. Furthermore, the proposed scheme provides more social welfare.

4.4 Summary

In this chapter, we have proposed two incentive mechanisms, such as Stackelberg game-based incentive mechanism and the auction theory-based incentive mechanism, for federated learning. In the first part, we have designed and analyzed a novel crowdsourcing framework to enable federated learning. An incentive mechanism has been established to enable the participation of several devices in federated learning. In particular, we have adopted a two-stage Stackelberg game model to jointly study the utility maximization of the participating clients and edge computing server interacting via an application platform for building a high-quality learning model. We have incorporated the challenge of maintaining communication efficiency for exchanging the model parameters among participating clients during aggregation. Further, we have derived the best response solution and proved the existence of Stackelberg equilibrium. We have examined the characteristics of participating clients for different parametric configurations. Additionally, we have conducted numerical simulations and presented several case studies to evaluate the framework's efficacy. Through a probabilistic model, we have designed and presented numerical results on an admission control strategy for the number of client's participation to attain the corresponding local consensus accuracy. In the second part, we formulated the incentive problem between the BS and mobile users in the federated learning service market as the auction game with the objective of maximizing social welfare. Then, we presented the method for mobile users to decide the bids submitted to the BS so that mobile users can minimize the energy cost. We also proposed the iterative algorithm with low complexity. In addition, we proposed a primal-dual greedy algorithm to tackle the NP-hard winner selection problem. Finally, we showed that the proposed auction mechanism guarantees truthfulness, individual rationality, and computation efficiency. Simulation results demonstrated the effectiveness of the proposed mechanism where social welfare obtained by our proposed mechanism is 400% larger than by the fixed price scheme. The model in our work can be extended to multi BS when users are one a large area. One BS can not cover the whole area. In that case, one BS performs edge aggregations of local models which are transmitted from devices in proximity. When each BS achieves a given learning accuracy, updated models at the edge are transmitted to the cloud or macro base station for global aggregation. Intuitively, this hierarchical model can help to reduce significant communication overhead between device users and the cloud via edge model aggregations and reduce the latency. In addition, through the coordination by the edge servers in proximity, more efficient communication and computation resource allocation among device users can be achieved. Moreover, we can consider the hierarchical auction mechanism consisting of two hierarchical auction models. i.e. a single-seller multiple-buyer model where the lower stage is between BS and mobile users and the higher stage is between the cloud and base stations. Another direction is the case which there are many base stations from different organizers who are interested in using the data from the set of users to train similar types of machine learning models. In that situation, there may be a competition of base stations. This will make base stations' decision-making different from our work. Therefore, we can also consider it as future work.

Appendix

A.1　KKT Solution

The utility maximization problem in (4.21) is a convex optimization problem whose optimal solution can be obtained by using Lagrangian duality. The Lagrangian of (4.21) is

$$\mathcal{L}(r, x(\epsilon), \lambda) = \beta \left(1 - 10^{-(ax(\epsilon)+b)} \right) - r \sum_{k \in K} (1 - \theta_k^*(r))$$
$$+ \lambda \left[\delta(1 - \max_k \theta_k^*(r) - x(\epsilon) \right] \tag{A.1}$$

where $\lambda \geq 0$ is the Lagrangian multiplier for constraint (4.22).

By taking the first-order derivative of (A.1) with respect to $x(\epsilon)$ and λ, KKT conditions are expressed as follows:

$$\frac{\partial \mathcal{L}}{\partial x(\epsilon)} = a\beta e^{-(a(x(\epsilon))+b)} - \lambda \leq 0, \text{ if } x(\epsilon) \geq 0. \tag{A.2}$$

$$\frac{\partial \mathcal{L}}{\partial \lambda} = \left[\delta(1 - \max_k \theta_k^*(r)) - x(\epsilon) \right] \geq 0, \text{ if } \lambda \geq 0. \tag{A.3}$$

By solving (A.2), the solution to the utility maximization problem (4.21) is

$$x^*(\epsilon) = \frac{-\ln(\lambda/a\beta) - b}{a}. \tag{A.4}$$

From (A.3), the Lagrangian multiplier λ is as

$$\lambda^* = a\beta e^{[a\delta(1 - \max_k \theta_k^*(r))+b]}. \tag{A.5}$$

Thus, from (A.4) and (A.5) the optimal solution to the utility maximization problem (4.21) is

$$x^*(\epsilon) = \delta(1 - \max_k \theta_k^*(r)). \tag{A.6}$$

Chapter 5
Security and Privacy

Abstract Federated learning allows data to be locally trained in their device and only send model updates to the central server for aggregation. But the security of model updates in the aggregation should also be carefully addressed. Existing works mainly focus on secure multiparty computation or differential privacy, which depends on heavy encryption or brings low accuracy. In this chapter, we discuss an efficient secure aggregation method for model updates in federated learning by pre-processing the model updates from each participant and only encrypting portion of the processed updates by functional encryption for inner product to protect the whole parameters, thus achieving efficient aggregation of model update vectors.

5.1 Introduction

The emerging Internet of Things (IoT) has connected a large volume of IoT devices, which will produce a vast amount of data in edge networks. According to the survey [139] by Statista, the total data volume of connected IoT devices worldwide is forecast to reach 79.4 zettabytes (ZBs) by 2025. The massive data will provide more possibility and chance for research and applications. For example, machine learning algorithms [140] will consume and dispose huge volumes of data to learn complex patterns about people and events to make predictions, which relies on the availability of an enormous amount data for training.

Traditional machine learning needs the data to be centrally trained in center servers. However, in real life, the data are scattered across different devices and organizations and cannot be easily integrated under practical constraints [141]. For instance, in different commercial banks, rather than each bank creating their own predictive models to predict the economic trends, each bank wants to establish a model learned over the whole statement. Nevertheless, the users' privacy does not allow them to share the data to others to construct a global predictive model.

Federated learning (FL) [142] can address these privacy issues by allowing participants to locally train the model using the local computing resources, and the participants only need to send the model updates to the central server for global model update. The central server, called aggregator, will merge the model

C. S. Hong et al., *Federated Learning for Wireless Networks*, Wireless Networks,
https://doi.org/10.1007/978-981-16-4963-9_5

parameters from different participants and send the aggregated parameters back to the participants for the next epoch training. The implementation of FL for model training at local participants can efficiently use the bandwidth of the network and protect the privacy of the raw data of participants. However, FL also faces some security risks, such as inference attack [143] and inversion attack [144], which can derive private information from local model updates. Therefore, it is necessary to design a secure aggregation scheme for the server to compute the weighted sum of the model updates from participants without leaking certain users' privacy.

To tackle such security issues and realize the secure aggregation of the model updates from participants, some work utilizes differential privacy (DP) [22, 145] to prevent the FL server from identifying the owner of a local update. The DP in FL will add certain noises to the local model updates, which may lower the performance of the resulting model. Some other methods [146] are based on secure multiparty computation (SMC). SMC computes a multiparty sum where no participant reveals its update clearly, which is mainly based on the cryptography methods, such as secret sharing, additive masking [147] and homomorphic encryption [148, 149] (e.g., the Paillier cryptosystem). These mechanisms may cause large encryption overhead and high data transmission cost. Meanwhile, there are also some hybrid methods [150, 151] combining the differential privacy and the SMC protocol to protect the model updates in FL.

This chapter aims to design an efficient secure aggregation scheme for model update in FL. The key insights are two-fold: First, the model updates of each participant are protected by matrix transformation, all or nothing transform. In this manner, the attacker cannot learn anything about the model updates without having obtained all the information of the transformed results. So we can protect the security of model updates by encrypting only a small portion of the transformed results. Second, we use the multi-input functional encryption for inner product to encrypt a small portion of transformed model updates, while the aggregator can precisely decrypt the aggregated encrypted portion of model update from the inner product, thus recovering aggregated model update vector with the unencrypted portion. Our design well suits the features of functional encryption for inner product and secure aggregation. Meanwhile, the linear transformation and encryption are quite efficient, which require small amount of computation.

5.2 Functional Encryption Enabled Federated Learning

5.2.1 Federated Learning

Federated learning [16] is a learning process in which the data owners collaboratively train a model, in which process any data owner does not expose its data to others.

We assume a standard federated learning, in which data is distributed across multiple participants P^j and cannot be shared. We use the neural network as the underlying machine learning model. The distributed learning process is performed in synchronous update rounds over a set of participants, in which a weighted average of the j participant model updates, with each weight set as η_j, is applied to the model.

At the participant side, each participant performs the local training based on the loss function, and transmit the model updates $\mathbf{W^j}$ to the aggregator.

At the aggregator side, the local model update parameters are aggregated, and the global model parameters $\omega_{g,t}$ are updated at each global aggregation iteration t as follows:

$$\omega_{g,t+1} = \omega_{g,t} + \sum_j \eta_j \mathbf{W^{j,t}} \tag{5.1}$$

where $\mathbf{W}^{j,t}$ is the model parameter update for each participant j at iteration t. The FL training process is iterated till the global loss function converges, or a desirable accuracy is achieved.

The secure aggregation in FL involves high-dimensional vectors $\mathbf{W}^{j,t}$ based on the training results from the participants' data. However, the aggregator does not need to get the individual's updates but to compute the element-wise weighted sum $\sum_j \eta_j \cdot \mathbf{W^{j,t}}$ of these update vectors. Hence, FL needs a secure aggregation scheme to compute the weighted sum while ensuring that the single user's privacy will not be leaked. There is also a need for FL to process high dimensional vectors and be communication efficient.

5.2.2   All or Nothing Transform (AONT)

AONT [152] transforms data into the encoded format, and it is hard to invert the encoded format back to the original data unless all of the encoded output is known. Stinson [153] defines linear AONT as follows, which can maintain the property of AONT while reducing the computational complexity.

Given a positive integer n, a finite field $\mathbb{F}_q$ with order q, a function π which maps an input of n-tuple $(x_1, \cdots, x_n)$ to an output of n-tuple, i.e., $(y_1, \ldots, y_n)$, where $x_i, y_i \in \mathbb{F}_q$ and $1 \le i \le n$, we say π is a linear $(n, q) - AONT$, if it satisfies the following conditions:

- π is a bijection;
- Each y_i $(1 \le i \le n)$ is an $\mathbb{F}_q$-linear function of $x_1, \ldots, x_i, \ldots, x_n$ $(1 \le i \le n)$;
- If any $n - 1$ out of n output values $y_1, \ldots, y_i, \ldots, y_n$ are fixed, any input value x_i $(1 \le i \le n)$ is completely undetermined.

An $n \times n$ encoding matrix for the linear (n, q)-AONT can be constructed as [153]:

$$
\mathbf{M} = \begin{pmatrix} 1 & 0 & \cdots & 0 & 1 \\ 0 & 1 & \cdots & 0 & 1 \\ \vdots & \vdots & \ddots & \vdots & \vdots \\ 0 & 0 & \cdots & 1 & 1 \\ 1 & 1 & \cdots & 1 & \lambda \end{pmatrix}.
$$

Each element in $\mathbf{M}$ is chosen from the finite field $\mathbb{F}_q$, where $q = p^k$, and p is a prime number and k is a positive integer. $\lambda \in \mathbb{F}_q$ such that

$$
\lambda \notin \{(n-1) \bmod p, (n-2) \bmod p\}. \tag{5.2}
$$

5.2.3   Multi-Input Functional Encryption for Inner Product

Functional encryption [154] allows users to learn specific functions of the encrypted data without learning any information about the plain text. Both functional encryption (FE) and homomorphic encryption (HE) can compute over encrypted data, but the difference between FE and HE is that FE can directly learn a function of what the ciphertext is encrypting, while HE also needs to decrypt the ciphertext to get the computed result.

For any function f from a class $\mathbf{F}$, a functional decryption key sk_f can be computed such that, given any ciphertext c with underlying plaintext x, using sk_f, a user can efficiently compute $f(x)$, but does not get any additional information about plaintext x.

However, the input can be large, the functional encryption can be extended to multi-input functional encryption (MIFE). For example, there are n participants which can generate ciphertexts $En(x_1), En(x_2), \cdots, En(x_n)$, one can use a secret key sk_f to retrieve $f(x_1, \cdots, x_n)$ without knowing any plaintext x_i.

Although there are several different types of functional encryption, such as identity-based encryption schemes, attribute-based encryption schemes and inner product functional encryption (IPFE) schemes, etc. We consider the MIFE schemes for the inner-product functionality where the plaintexts are vectors and the encrypted data can be used along with an evaluation key to compute the inner product of the said vectors with weight vector $\mathbf{y} = (y_1, y_2, \cdots, y_n)$, as shown in Fig. 5.1.

We choose IPFE for the reason that the weighted aggregation process of the model updates from each participant in FL at the aggregator is similar to the calculation of the inner product between input vectors $\mathbf{x}$ and $\mathbf{y}$. Also the IPFE is efficient due to the linear encryption. But there exists a problem that the result of the inner product is a scalar, while the aggregated result of model updates is a vector. To address this problem, before each participant performs the IPFE, we will apply an all or nothing transform described in Sect. 5.2.2 to the model updates, thus the

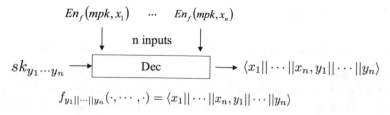

$$f_{y_1||\cdots||y_n}(\cdot, \cdots, \cdot) = \langle x_1||\cdots||x_n, y_1||\cdots||y_n\rangle$$

Fig. 5.1 Multi-input inner product

vector can be protected by partially encrypted. Then the aggregator can recover the aggregated vectors from the inner product and the transformation results. We now present our construction of a MIFE scheme for FL, which is derived from the construction in [155]. Suppose the inner product function defined by the weight vector $\mathbf{y}$ is $f_{\mathbf{y}}(\mathbf{x}) = \sum_{i=1}^{n}(x_i \cdot y_i)$, where n denotes the total number of input sources. The construction of Multi-input Functional Encryption for Inner Product is defined as follows:

1. SetUp(λ): Takes as input the security parameter, and generates prime-order group $\mathcal{G} := (\mathbb{G}, p, g) \leftarrow GGen(1^\lambda)$, and $\mathcal{H}$ a full domain function onto $\mathbb{G}^2$. It also generates the encryption keys $\mathbf{s}_i \leftarrow \mathbb{Z}_p^2$, for $i = 1, 2, \cdots, n$. Then, the public parameters master public key mpk consist of $(\mathcal{G}, p, g, \mathcal{H})$. The encryption keys are $\mathbf{ek}_i = \mathbf{s}_i$ for $i = 1, 2, \cdots, n$, and the master secret key is $msk = ((\mathbf{ek}_i)_i)$.
2. Key Distribute($\mathcal{G}, \mathbf{ek}_i$): The Trusted Party will distribute the encryption keys to each participant.
3. Encrypt($\mathbf{ek}_i, x_i, \ell$): Each participant takes the value x_i as input to encrypt, under the key $\mathbf{ek}_i = \mathbf{s}_i$ and the label ℓ. It computes $[\mathbf{u}_\ell] := \mathcal{H}(\ell) \in \mathbb{G}^2$, and outputs the ciphertext $[c_i] = [\mathbf{u}_\ell^T \mathbf{s}_i + x_i] \in \mathbb{G}$;
4. Decrypt Key Generate(msk,($\mathbf{y}$)): The Trusted Party takes the $msk = (\mathbf{s}_i)_i$ and the weight vector $\mathbf{y}$ as input, and outputs the functional decryption key $dk_{\mathbf{y}} = (\mathbf{y}, \mathbf{d}) = (\mathbf{y}, \sum_i \mathbf{s}_i \cdot y_i) \in \mathbb{Z}_p^n \times \mathbb{Z}_p^2$;
5. Decrypt($dk_{\mathbf{y}}, \ell, ([c_i])_{i\in[n]}$): The aggregator takes a functional decryption key $dk_{\mathbf{y}} = (\mathbf{y}, \mathbf{d})$, a label ℓ, and ciphertexts as inputs. It then computes $[\mathbf{u}_\ell] := \mathcal{H}(\ell)$, $[\alpha] = \sum_i [c_i] \cdot y_i - [\mathbf{u}_\ell] \cdot \mathbf{d}$, and eventually solves the discrete logarithm to extract and return α.

Let $\mathbf{x}, \mathbf{y} \in \mathbb{Z}_p^n$, we can calculate:

$[\alpha] = \sum_i [c_i] \cdot y_i - [\mathbf{u}_\ell^T] \cdot \mathbf{d} = \sum_i [\mathbf{u}_\ell^T \mathbf{s}_i + x_i] \cdot \mathbf{y}_i - [\mathbf{u}_\ell^T] \cdot \sum_i y_i \mathbf{s}_i = \sum_i [\mathbf{u}_\ell^T] \cdot$
$\mathbf{s}_i y_i + \sum_i [x_i] \cdot y_i - [\mathbf{u}_\ell^T] \cdot \sum_i y_i \mathbf{x}_i = [\sum_i s_i y_i]$, which well suits the pre-defined inner product function for weight vector $\mathbf{y}$: $f_{\mathbf{y}}(\mathbf{x}) = \sum_{i=1}^{n}(x_i \cdot y_i)$.

5.2.4 Threat Model

Here, we mainly consider about two kinds of adversaries in our scheme:

- **Outside adversaries**. During transmission, the model parameters of the participant may be eavesdropped by the outside adversaries. Although the model updates are vectors, others can still perform inference attack to learn individual's private information.
- **Inside adversaries**. In FL, the curious aggregator may try to obtain the model update parameters of the individual through the process of secure aggregation, and participants may conduct colluding attack with the aggregator to learn about the single participant's private parameters.
- **Assumptions**. We assume that there is a secure key distribution protocol for trusted party to distribute keys to the aggregator and participants. We also assume that the communication between participants and aggregator is secure, for the transmitted data cannot be modified. Among the total number of n participants, we assume that at least the number of d participants are non-colluding participants.

5.3   Secure Aggregation for Wireless Federated Learning

We suppose that each participant P^j holds its own data D^j. When participant receives the query Q from the aggregator in one epoch, by training the local data, the participant will get the model update $\mathbf{W^j}$ and send it to the aggregator $\mathcal{A}$.

To protect the model updates $\mathbf{W^j}$ from each participant P^j, we will apply the matrix transform AONT and multi-input functional encryption for inner product. The traditional multi-input functional encryption will only output the inner product of input vector and the weight vector, and the result of inner product is a scalar. But in FL, the result of secure aggregation is a high-dimensional vector for machine learning. To solve this problem, we apply the AONT to protect model updates $\mathbf{W^j}$. The AONT transform can ensure that by encrypting partial information of transformed $\mathbf{W^j}$, the security of the whole $\mathbf{W^j}$ can be guaranteed. So we only need to put the partial information of transformed $\mathbf{W^j}$ from multi-participant as the encrypted element of the input vector of functional encryption, and then set the weight of each participant's model as the weight vector of functional encryption. Afterwards, the encrypted portion and unencrypted portion of the transformed $\mathbf{W^j}$ can be weightedly aggregated separately at aggregator. At last, the aggregator can perform the reverse AONT on the weighted sum to recover the element-wise aggregated model update vector of the participants. The process is described in Algorithm 10. We will show it detailedly below.

5.3.1 Participant Pre-processing Mode Updates

After the participants receive the training request from the server, they will train their local database to minimize the loss function. Then the local model updates of each participant will be obtained as a response to the aggregator. Before sending the model updates to the aggregator, some preprocessing will be conducted to the model updates to protect the security of these parameters:

First, we use all or nothing transform (AONT) to preprocess the model updates from each participant. For model updates $\mathbf{W^j}$ of participant P^j, it can be depicted as: $\mathbf{W}^j = \left(w_1^j, w_2^j, \cdots, w_m^j\right)^T$. Use linear (m, q)-AONT matrix described in Sect. 5.2.2 to process the message we can get $\mathbf{U}^j = \mathbf{M} \cdot \mathbf{W}^j$:

$$\mathbf{U}^j = \begin{cases} u_1^j = w_1^j + w_m^j \\ u_2^j = w_2^j + w_m^j \\ \cdots\cdots \\ u_{m-1}^j = w_{m-1}^j + w_m^j \\ u_m^j = w_1^j + w_2^j + \cdots + w_{m-1}^j + \lambda \cdot w_m^j. \end{cases} \tag{5.3}$$

AONT, known as all-or-nothing protocol, is an encryption mode which allows the data to be understood only if all of it is known. We use the linear AONT to reduce the computation complexity while maintaining the property of AONT. Therefore, after being processed by linear (m, q)-AONT, others cannot learn anything in W_i^j, unless they know all the information in $\mathbf{U}^j$.

The inverse transform of Eq. (5.3) to recover the original model updates W^j is:

$$\mathbf{W}^j = \begin{cases} w_i^j = u_i^j - \gamma \cdot \left(u_1^j + u_2^j + \cdots + u_{m-1}^j - u_m^j\right), \\ \qquad\qquad for \quad 1 \le i \le m - 1 \\ w_m^j = \gamma \left(u_1^j + u_2^j + \cdots + u_{m-1}^j - u_m^j\right), \end{cases} \tag{5.4}$$

where $\gamma = (m - 1 - \lambda)^{-1}$.

We can see from the inverse transform shown in Eq. (5.4) that the reconstruction of each element of the model updates $\mathbf{W}^j$ is associated with each element in $\mathbf{U}^j$. Then we can encrypt the element u_m^j in $\mathbf{U}^j$ as one input of multi-input functional encryption mentioned in Sect. 5.2.3 with the key distributed by trusted party.

Algorithm 10 Secure federated learning

1: **Input:** $\mathcal{L}_{FL} :=$ Machine learning algorithms to be trained; $S_P :=$ set of participants; $\mathbf{W}^j :=$ the model updates from participant j; $\eta_j :=$ the weight for $\mathbf{W}^j$; $n - d + 1 :=$ minimum number of aggregated replies;
2: **Output:** Trained global model $\mathbf{G}$;
3: **for each** $P^j \in S_P$ **do**
4:     $\mathcal{A}$ queries P^j with $\mathcal{L}_{FL}$
5:     P^j trains local data and obtains model updates $\mathbf{W}^j$
6:     P^j performs AONT transform: $\mathbf{U}^j = \mathbf{M} \cdot \mathbf{W}^j$
7:     P^j partially encrypts $En_{fe}(u_m^j)$ in $\mathbf{U}^j$ and sends $\mathbf{U}^j$ to $\mathcal{A}$
8: **end for**
9: $\mathcal{A}$ aggregates at least $n - d + 1$ encrypted model updates $\mathbf{U^j}$;
10: $\mathcal{A}$ requires decryprtion key based on weight vector $\boldsymbol{\eta}$;
11: $\mathcal{A}$ computes $\sum_j \eta_j \mathbf{W^j}$ from aggregated $\mathbf{U^j}$;
12: $\mathcal{A}$ updates $\mathbf{G}$ with $\sum_j \eta_j \mathbf{W^j}$;
13: **Return G**;

The partially encrypted $\mathbf{U}^j$ can be depicted as

$$\mathbf{U^j} = \begin{pmatrix} u_1^j \\ u_2^j \\ \cdots \\ u_{m-1}^j \\ u_m^j \end{pmatrix} \rightarrow \begin{pmatrix} u_1^j \\ u_2^j \\ \cdots \\ u_{m-1}^j \\ En_f^j\left(u_m^j\right) \end{pmatrix}. \tag{5.5}$$

Each participant P^j will send the transformed and partially encrypted model updates $\mathbf{U^j}$ to the aggregator. We assume that there are n participants with model updates: $\mathbf{U}^1, \mathbf{U}^2, \cdots, \mathbf{U^j}, \cdots, \mathbf{U^n}$. The encrypted multi-input vector of functional encryption can be depicted as $En_f^1\left(u_m^1\right), En_f^2\left(u_m^2\right), \cdots,$ $En_f^j\left(u_m^j\right), \cdots, En_f^n\left(u_m^n\right)$.

5.3.2   Secure Aggregation at Aggregator

Because we assume that at least d participants are non-colluding, after the aggregator collects at least $n - d + 1$ model updates $\mathbf{U}^j$ from participants, it can aggregate and decrypt the model updates to update the global model and send updated global model parameters to the participants to support next epoch training.

The aggregator receives a set of $\mathbb{U}$ model updates from participants S_P^{Thres}, where $|S_P^{Thres}| = p \geq n - d + 1$. For each update $\mathbf{U^j}$ in $\mathbb{U}$ ($1 \leq j \leq p$), we first need to aggregate and decrypt the encrypted u_m^j:

$$
\begin{aligned}
&En_f^1\left(u_m^1\right), En_f^2\left(u_m^2\right), \cdots, En_f^j\left(u_m^j\right), \cdots, En_f^p\left(u_m^p\right) \\
&\xrightarrow{\text{Aggregate}} \quad f_\eta\left(u_m^1, u_m^2, \cdots, u_m^p\right) \\
&= \left\langle u_m^1 || \cdots || u_m^p, \eta_1 || \cdots || \eta_p \right\rangle \\
&= \sum_i u_m^i \cdot \eta_i
\end{aligned}
\tag{5.6}
$$

where $\eta = \left(\eta_1, \eta_2, \cdots, \eta_p\right)$ is the weight vector for aggregation according to the training model or set by the aggregator.

At last, the aggregator can aggregate with the unencrypted portion $\left(u_1^j, u_2^j, \cdots, u_{m-1}^j\right)$ ($1 \leq j \leq p$) and the decrypted aggregated portion $f_\eta\left(u_m^1, u_m^2, \cdots, u_m^p\right)$ from p participants to recover the final weighted aggregated model updates $\sum_j \eta_j \mathbf{W^j} = \left(\sum_j \eta_j w_1^j, \sum_j \eta_j w_2^j, \cdots, \sum_j \eta_j w_m^j\right)$:

$$
\begin{cases}
\sum_{j=1}^p \eta_j w_i^j = \sum_{j=1}^p \eta_j u_i^j - \gamma \sum_{j=1}^p \eta_j \sum_{k=1}^{m-1} u_k^j \\
\qquad + \gamma f_{\mathbf{y}}\left(u_m^1, u_m^2, \cdots, u_m^p\right) \quad (1 \leq i \leq m-1) \\
\sum_{j=1}^p \eta_j w_m^j = \gamma \left(\sum_{j=1}^p \eta_j \sum_{k=1}^{m-1} u_k^j - f_{\mathbf{y}}\left(u_m^1, u_m^2, \cdots, u_m^p\right)\right).
\end{cases}
\tag{5.7}
$$

As we can see from Eq. (5.7), the aggregator can reconstruct the weighted sum of each participant's model updates without knowing the information of the individual due to the features of AONT and multi-input functional encryption. Because the aggregator can only learn the weighted sum of u_m^j from each participant, and without this part, the aggregator cannot learn anything about the original model updates $\mathbf{U^j}$. Meanwhile, the small amount of encryption and the linear operation ensure the efficiency of our scheme. At last, the whole progress procedure of one epoch in federated learning is shown in Fig. 5.2.

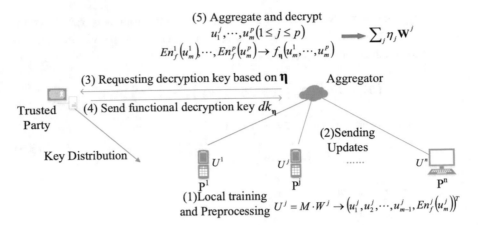

Fig. 5.2 Framework of Secure Aggregation in one epoch

5.4   Security Analysis

5.4.1   Security for Encryption

In each participant, linear AONT and functional encryption are utilized to protect the model update vector. We will first demonstrate that by applying linear AONT transformation to the vector, others must obtain the whole transformed vector to obtain the original vector. So we can encrypt partial information of the transformed vector to protect the whole original vector. Then we will demonstrate that the multi-input functional encryption [155] chosen to partially encrypt the transformed vector can ensure the security of the data.

First, we prove the security of linear AONT. For a linear $(n, q) - AONT$ [153], the input vector $V = (v_1 \; v_2 \; \cdots \; v_n)^T$ and the resulting vector $U = (u_1 \; u_2 \; \cdots \; u_n)^T$, from $U = M \cdot V$, we can get:

$$\begin{cases} u_i = v_i + v_n, \; for \; i = 1, 2, \cdots, n - 1 \\ u_n = v_1 + v_2 + \cdots + \lambda v_n. \end{cases} \tag{5.8}$$

Inversely, by analyzing output vector $U = (u_1 \; u_2 \; \cdots \; u_n)^T$, we can obtain the input vector $V = (v_1 \; v_2 \; \cdots \; v_n)^T$:

$$\begin{cases} v_i = u_i - \gamma \, (u_1 + u_2 + \cdots + u_{n-1} - u_n), \; 1 \le i \le n - 1 \\ v_n = \gamma \, (u_1 + u_2 + \cdots + u_{n-1} - u_n), \end{cases} \tag{5.9}$$

where $\gamma = (n - 1 - \lambda)^{-1}$. From the aforementioned equations, we can see that in order to obtain the value of any one of the input vector $V = (v_1\ v_2\ \cdots\ v_n)^T$, all the generated transformed vector must be obtained.

Therefore, we can ensure the security of the input vector V by encrypt small portion of the transformed vector U. The encryption method we choose is multi-input functional encryption [154], which has been proved to be secure in terms of indistinguishability under the Decisional Diffie–Hellman assumption. What we have done to this method is to add a key distribution method for multi-participant in the FL to encrypt their model updates, which will not change the encryption security.

In conclusion, the encryption we use is secure. For the outside adversaries, although they can eavesdrop transmitted information, they still cannot learn any information of the original model updates from the transformed model updates.

5.4.2 Privacy for Participant

For curious participants who may conduct colluding attack with aggregator, we assume that at least d participants are non-colluding and the total number of participants is n. Therefore, the aggregator need to aggregate at least $n - d + 1$ model updates from the participant to update the global model to defend against colluding attack. Because the decryption key for the multi-input functional encryption is generated based on weight vector which is detailedly described in Sect. 5.2.3, we can set a threshold for trusted party who will only response the decryption key to the request in which the non-zero element in the weight vector is no less than $n - d + 1$. In this way, the scheme can defend against the participants colluding attack.

The curious aggregator may want to obtain the private information of some certain user. But the aggregator can only learn the aggregated value of at least $n - d + 1$ participants based on the setting above. Although the aggregator can learn some information about the transformed model updates U^j from individual participant, due to the feature of AONT, without knowing all the information about the transformed vector U^j, the curious aggregator cannot learn any information about the original model updates W^j of this participant.

5.5 Implementation and Evaluation

5.5.1 Implementation

We train a convolutional neural network (CNN) to classify the CIFAR10 dataset of images. The model contains convolutional layers, fully connected layer, max pooling layer, normalization layer and dropout layer with the cross entropy loss. We conduct the experiments with 100 participants, in each epoch, the aggregator randomly chooses a subset of participants to train the model with learning rate as 0.0001 and batch size as 32. Our scheme is implemented in Keras with Tensorflow backend.

The experiment is performed on a local machine, equipped with an Intel Core i7 and 16G RAM, running Ubuntu 16.04.

As for encryption based secure aggregation, we compare the efficiency of our functional encryption (FE) based scheme with the prevalent homomorphic encryption (HE) based scheme for federated learning. The baseline (Base) scheme is FL without the encryption method.

5.5.2   Evaluation

Figure 5.3 shows the accumulated training time as epoch increases in FL. We can see from this figure that our functional encryption based secure aggregation scheme (FE) does not bring to much time overhead compared to the unencrypted federated learning (Base). It also performs better than the homomorphic encryption based scheme. When the epoch is 60, the simulated training time of FE based scheme is about 3 min longer than the baseline, while that of HE based scheme is about 8 min longer than the baseline. This is because, the homomorphic encryption need to decrypt the ciphertexts after aggregated at the sever, while the functional encryption will directly obtain the weighted functional result of the ciphertexts from

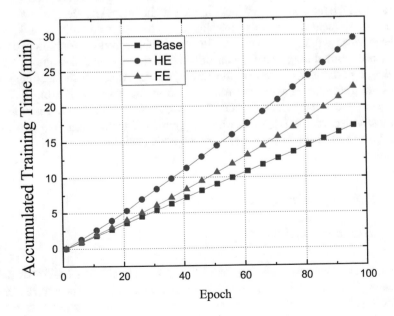

Fig. 5.3 Training time

participants. In our scheme, the aggregator will only additionally perform the inverse AONT which is a linear transformation to obtain the weighted aggregated model updates vector. Also, the homomorphic encryption itself consumes much time than functional encryption.

For the data to be transmitted during the FL progress, our scheme need to transmit the partially encrypted model updates from participants to the aggregator and send the weight vector to the Trusted Party to request the corresponding functional decryption key at each epoch (see Fig. 5.2), which will not increase the throughput of the network too much, thus saving bandwidth. Whereas in the scheme applying threshold homomorphic encryption, after the aggregator summed up at least $n-d+1$ model updates from participants, he/she will also need to send this aggregated encrypted model updates to each participant requesting partial decryption to defend against collusion attacks. Then after collecting the partially decrypted aggregated model updates, the aggregator can compute the final weighted summed updates. So the HE based scheme will have more $(n - d + 1) \times m$ data transmission than our scheme in each training epoch.

5.6 Summary

Currently, a large volume of IoT devices generate huge amounts of data in edge networks, which can open up many research and applications for machine learning. However, traditional machine learning requires data to be sent to a server and centrally trained, which will cause the waste of the bandwidth and expose privacy of individuals. Federated learning allows data to be locally trained in their device and only send model updates to the central server for aggregation. But the security of model updates in the aggregation should also be carefully addressed. Existing works mainly focus on secure multiparty computation or differential privacy, which depends on heavy encryption or brings low accuracy. In this chapter, we discuss an efficient secure aggregation scheme for model update in FL. We combine the all or nothing transform and multi-input functional encryption, which well suit the weighted aggregation feature of FL and solve the problem of vector aggregation of multi-input functional encryption. Meanwhile, the small amount of linear encryption and transformation will bring small overhead. Security analysis and experimental evaluation confirm that this scheme ensure the security of aggregation of FL with an acceptable overhead.

Chapter 6
Unsupervised Federated Learning

Abstract In this chapter, we consider unsupervised learning tasks being implemented within the federated learning framework to satisfy stringent requirements for low-latency and privacy of the emerging applications. The discussed algorithm is based on Dual Averaging (DA), where the gradients of each agent are aggregated at a central node. While having its advantages in terms of distributed computation, the accuracy of federated learning training reduces significantly when the data is nonuniformly distributed across devices. Therefore, this chapter discusses two weight computation algorithms, with one using a fixed size bin and the other with self-organizing maps (SOM) that solves the underlying dimensionality problem inherent in the first method.

6.1 Introduction

Traditionally, machine learning, which consists of training and inference phases, is done at the centralized cloud computing, which can sustain more heavy computations than edge devices. However, nowadays imbuing edge devices with intelligence gives ability to train models locally, while providing privacy, security, regulatory and economic benefits. Recently proposed federated learning [58] allows mobile edge devices, with limited computational resources, such as mobile phones and IoT devices, to learn a global model for prediction, while only perform training with local data [156]. Given the fact that prediction is done at the edge, stringent requirements in low-latency applications can be met.

Clustering is one of the fundamental tasks in data analysis and signal processing, and plays an important role in various applications such as social network analysis, image and video processing and autonomous driving. These tasks are usually carried out at central servers and require the collected data to be transmitted from individual nodes/agents. While previous works on federated learning have only considered supervised learning with neural networks, the goal of this work is to propose a framework that supports unsupervised learning under the federated learning architecture.

Despite its advantages, there are few challenges that must be overcome before federated learning can be deployed in a large scale. The first challenge is unpredictable users, or processing agents, behaviors such as the unreliable device connectivity, interrupted execution, difference in convergence time for local training, which was considered in [157–159]. The second problem is the imbalance in collected data across the network, i.e., particular sensors may have limited observations of certain events, which imposes significant statistical challenges in the learning process. Several works have analyzed and attempted to resolve the problem. Reference [54] has proposed to learn separate models for each node through a multi-task learning framework. However, this approach creates additional communication bandwidth overhead, which the authors attempt to resolve. Reference [160] has shown that accuracy of federated learning reduces significantly, by up to 55%, for neural networks trained for highly skewed non-IID data. To resolve this issue, small subsets of data are globally shared among all edge devices. Experimental results have shown that sharing only 5% of local data can increase prediction accuracy by 30%. However, the approach is not applicable for unsupervised learning due to the absence of data labels.

Therefore, the unsupervised learning scheme under the federated learning framework needs to be considered. Here, assume processing agents in the network store non-identically distributed data sets. The algorithm relies on modifying the dual averaging (DA) algorithm [161], which is based on weighted gradient aggregation at the central node. However, proper weights need to be determined to reflect the non-IID nature of the observed data at the agents. Two methods are given to overcome this problem. The first scheme uses fixed size bins over the data to determine how much data is in a certain bin. However, the number of weights that needs to be computed grows exponentially with respect to the dimension of the data. The second method which exploits the mapping property of self-organizing maps (SOM) is utilized to tackle the curse of dimensionality problem.

6.2   Problem Formulation

A typical federated learning procedure is shown in Fig. 6.1. At the initial step, a subset of agents is selected and a pretrained model is downloaded to the agents (top left). Next, the agents compute an updated model based on their local data (top right). After training, the model updates are sent from the selected agents back to the server (bottom right) and the server aggregates these models to construct an improved global model by simple averaging or other techniques (bottom left). This process continues until convergence or until the desired level of prediction accuracy is reached. Notice that in practical implementations of big models, whole model is not transmitted to the server but rather, only structural updates are sent.

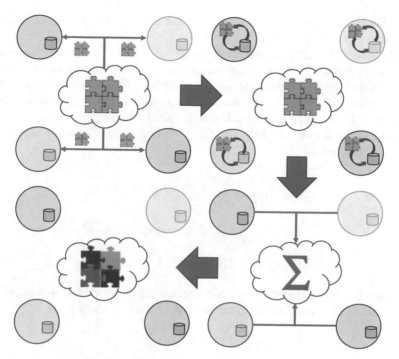

Fig. 6.1 Unsupervised federated learning procedure

Assume the network consists of a set $\mathcal{J} = \{1, \ldots, J\}$ of data collecting and processing agents.[1] The jth agent can send information about certain parameters to the centralized server. Parameter update is done at the server and is based on a convex combination of parameters that are received. To model this weighting process, let $\mathbf{w} \in \mathbb{R}^J$ be a vector containing nonnegative weights that is associated with how much emphasis should be placed on the estimated parameter for each agent, with $[\mathbf{w}]_j > 0$ denoting that a connection exists between the jth node and centralized server. It is assumed that $\mathbf{w}$ is a stationary stochastic vector, so that $\mathbf{w}^T \mathbf{1}_J = 1$. It is shown in Sect. 6.4 that proper weighting is crucial to guarantee good performance and proper determination of it is described in the sequel.

Assume that sensor node j observes a set of data $\mathcal{X}_j := \{\mathbf{x}_{jn}, n \in \mathcal{N}_j\}$ with $\mathcal{N}_j = \{1, \ldots, N_j\}$ being a set of neighboring nodes, where $\mathbf{x}_{jn} \in \mathbb{R}^q$ denotes the nth observation at the jth node and q denotes the data dimension. Each observation is assumed to be drawn from one class $\mathcal{C}_k$ with $k \in \{1, \ldots, K\} \triangleq \mathcal{K}$, where K denotes total number of classes and assumed to be known *a priori*, or can be

---

[1] Uppercase (lowercase) bold face letters indicate matrices (column vectors). Superscript ^{H} denotes Hermitian, ^{T} denotes transposition. $\mathbf{1}_M$ denotes an $M \times 1$ vector, containing 1 in all of its entries. $\langle \cdot, \cdot \rangle$ denotes the inner product operator. $\|\mathbf{a}\|$ denotes the ℓ_2 norm of $\mathbf{a}$.

estimated by various algorithms [162], e.g. the elbow method, average silhouette method or gap statistic method.

The goal is to assign each observation point to a particular cluster and estimate the cluster centroid. Presented below considers hard assignment scheme, i.e., each point only belongs to one cluster; however, it can easily be cast into a soft clustering scheme where each point is assigned to a cluster with a certain probability. Denote the centroid of cluster $\mathcal{C}_k$ as $\mathbf{m}_k \in \mathbb{R}^q$, and the membership label as $\mu_{jnk} \in \{0, 1\}$, so $\mu_{jnk} = 1$ if $\mathbf{x}_{jn}$ is assigned to $\mathcal{C}_k$ and $\mu_{jnk} = 0$, otherwise. The clustering problem can be formulated as

$$
\min_{\mu_{jnk},\mathbf{m}_k} \sum_{j \in \mathcal{J}} \sum_{k \in \mathcal{K}} \sum_{n \in \mathcal{N}_j} \frac{1}{2} \mu_{jnk} \|\mathbf{m}_k - \mathbf{x}_{jn}\|^2
$$

$$
\text{s.t.} \sum_{j \in \mathcal{J}} \mu_{jnk} = 1, \ k \in \mathcal{K}, n \in \mathcal{N}_j,
$$

$$
\mu_{jnk} \in \{0, 1\}, j \in \mathcal{J}, k \in \mathcal{K}, \ n \in \mathcal{N}_j,
$$

(6.1)

where the Euclidean distance between cluster centroids and corresponding assigned data points is minimized and the constraints describe hard assignment.

6.3   Dual Averaging Algorithm

In this section, dual averaging algorithm is described. In addition, two methods for weights computation, i.e. bin methods and self-organizing maps, are discussed as well.

6.3.1   Algorithm Description

Prior to clustering, all agents conduct simple data analysis and send statistics to server, that decides which weight each agent will have in a certain region of the data space. Next, network runs an iterative algorithm similar to the k-means algorithm, which consists of the following steps. First, cluster initialization using k-means++ [163] is done. This is followed by an iterative algorithm consisting of two steps. The first step is data labeling, where data points are assigned to specific cluster at each node. Inspired by simulated annealing, the proposed algorithm uses distance perturbation, which may avoid local minima and results in better clustering performance. The second step computes data centroids via the proposed DA based method, where weighted gradients are combined at the central node. Each step is described in the sequel.

6.3.2 Data Labeling Step

A simple way to solve (6.1) for a suboptimal μ_{jnk} is to employ the following procedure. After all agents have updated their respective centroids, each data point from the locally observed dataset at each agent is assigned to a specific cluster based on the distance from that agent's centroid, i.e. $\mu_{jnk} = 1$ if $k = \arg\min \|x_{jn} - m_{jk}^{(t_1)}\|$ and $\mu_{jnk} = 0$ otherwise. However, inspired by Selim and Alsultan [164], it is modified as

$$\mu_{jnk} = \begin{cases} 1, \text{ if } k = \arg\min(1 + \frac{\xi}{t_1})\|x_{jn} - m_k^{(t_1)}\|, \\ 0, \text{ otherwise}, \end{cases} \tag{6.2}$$

where ξ is random variable drawn from uniform distribution $\xi \sim \mathcal{U}(0; \xi_{max})$ between 0 and ξ_{max}. $\frac{\xi}{t_1}\|x_{jn} - m_{jk}^{(t_1)}\|$ in (6.2) can be interpreted as a perturbation term on the distance between centroid m_{jk} and point x_{jn}. Setting ξ_{max} to a large value leads to stronger perturbation and as a result, there is a higher chance to avoid local minima in (6.1), but at the cost of requiring more iterations for the algorithm to converge.

6.3.3 DA-Based Centroid Computation Step

A modified DA subgradient based method is introduced herein. Given μ_{jnk} from step (6.2), at (inner iteration) time step t_2 of the algorithm, each node calculates a gradient $g_{jk}^{(t_2)}$ of local objective function $\sum_{n \in \mathcal{N}_j} \frac{1}{2}\|m_k - x_{jn}\|^2$ as

$$g_{jk}^{(t_2)} = \sum_{n \in \mathcal{N}_j} \mu_{jnk}(m_k^{(t_2)} - x_{jn}). \tag{6.3}$$

Next, the DA generates a sequence of iterates $\{m_k^{(t_2)}, z_{jk}^{(t_2)}\}_{t_2=1}^{\infty}$ at the central node using the update equations

$$z_k^{(t_2+1)} = z_k^{(t_2)} + \sum_{j \in \mathcal{J}} [w]_j g_{jk}^{(t_2)}, \tag{6.4a}$$

$$m_k^{(t_2+1)} = \arg\min \langle z_k^{(t_2+1)}, m_k \rangle + \alpha^{(t_2)}\|m_k\|^2, \tag{6.4b}$$

where $z_k^{(t_2)}$ can be seen as accumulated gradient for cluster k at iteration t_2. First term on the right-hand side of (6.4b) is the first-order approximation of the objective and the second is a regularization term to turn the problem into strictly convex form, which prevents the solution from being unbounded. $\alpha^{(t_2)}$ is nonincreasing positive sequence and acts as regularization parameter and has to be carefully selected.

Notice that (6.4a) is different from that of the conventional DA algorithm in which weighting of $\mathbf{g}_{jk}^{(t_2)}$ is now performed. The proposed method can also be viewed as a combination of stochastic gradient descent (SGD) with minibatch gradient aggregation [165]. Similar to the latter, the data at each agent can be viewed as randomly selected data subset and the gradient is computed based on this subset. Hence the data at the agent is similar to a minibatch. In addition, the gradients from different agents are convexly combined at the central node during the centroid update step, which is the main difference between the proposed method and existing techniques. A data driven approach is used to compute the weights used in the above convex combination and it is described in the following subsection.

6.3.4   Weight Computation via Bin Method

Consider the scenario shown in Fig. 6.2, where 3 sensor nodes observe data points from 16 clusters indicated by different colors. It is clear that each agent in the figure only has partial observations of the entire dataset. It is intuitive to weigh the gradients $\mathbf{g}_{jk}^{(t_2)}$, $\forall j$ that correspond to each cluster by the number of points that are observed at agent j. In particular, it can be seen that agent 1 will have the highest weight for the cluster at the bottom left of the grid in Fig. 6.2, while agents 2 and 3 will have zero weight as this cluster is not observed by those agents. Unfortunately, weights cannot be computed in this fashion because the data labels remain unknown before clustering is done. Therefore, it is proposed to assign the weights by dividing the data space into a grid with uniform-sized bins as shown in Fig. 6.2 and calculate the number of points falling into a particular *bin* (a region of the grid) at each node. The number of points for each bin is sent to the central server from each agent so that the weight for each particular bin at each agent can be computed as the number of points falling in that bin at that agent divided by the total number of points in that bin from all agents. This guarantees that $\mathbf{1}_J^T \mathbf{w} = 1$ so that the algorithm will converge, which will be shown in later publication. The above approach, however,

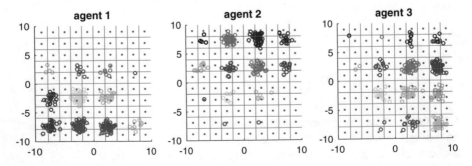

Fig. 6.2 Unbalanced data sets observed at different agents

suffers from the curse of dimensionality. For instance, in Fig. 6.2, the data dimension $q = 2$ and the number of bins required equals 10^q.

6.3.5   Weight Computation via Self-Organizing Maps

Self-organizing maps (SOM) [166], which is often used to produce a low-dimensional representation of the input space of training samples, can be used to resolve the curse of dimensionality problem. SOM consists of set of neurons $\mathcal{M}$ with its coordinate vector $\mathbf{p}_m \in \mathbb{R}^q, m \in \mathcal{M} = \{1, \dots, M\}$. Hence, the entire SOM is parameterized by $\mathbf{P}_j \triangleq [\mathbf{p}_1, \cdots, \mathbf{p}_M]$. Typically, training process consists of the following steps. (0) All neurons are initialized with small values of their weights. (1) For each data point, the neurons compute the distance to the data point and the closest neuron is declared as the winner. (2) The winning neuron determines the neighborhood of excited neurons and these neurons adjust their individual weights towards the data point. (3) Neurons decrease neighborhood radius and learning rate. These steps are repeated at each sensor until convergence and data map is obtained. Example of evolution of SOM training is shown on Fig. 6.3.

After $\mathbf{P}_j$ has been constructed from the local data observed at the jth agent, the proposed algorithm will send this back to the central server to form the aggreated SOM $\mathbf{P} \triangleq [\mathbf{P}_1, \cdots, \mathbf{P}_j]$. $\mathbf{P}$ will then be sent back to each node and each vector inside $\mathbf{P}$ will be used as a centroid for the bin similar to the one discussed in Sect. 6.3.4. The number of points falling into each bin is then calculated and the corresponding weights for the gradients in (6.4a) can be found. The overall scheme is summarized in Algorithm 11. Due to space limitation, only a brief discussion concerning the convergence of the proposed algorithm is given herein. The convergence rate of the centroid computation step (inner loop) is $O(\frac{\gamma}{T})$, where γ is a parameter that depends on the regularization parameter $\alpha^{(t_2)}$ and Euclidean distance between the initialized and optimal variable. T denotes the total number of iterations in the inner loop. Global convergence of the algorithm (outer loop) is similar to those of standard k-means, and probabilistic convergence upper bound can be computed based on [167, 168].

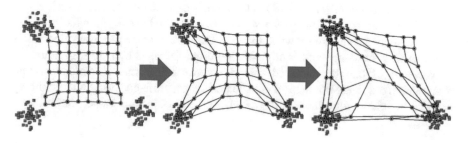

Fig. 6.3 Self-organizing maps training

Algorithm 11 DA-based unsupervised federated learning

1: **Input:** μ_{jnk}, $\mathbf{m}_k$, $k \in \mathcal{K}$, $j \in \mathcal{J}$, $n = 1 \ldots N_j$;

2: Each node j : Train SOM $\mathbf{P}_j$ and send it to server;

3: Server combines received models and send it to nodes;

4: Each node calculates number of points falling in the bin and sent it to server to determine gradient weight $\mathbf{w}$;

5: Initialize $\mathbf{m}_k^{(0)}$ via k-means++, send it to nodes; $t_1 = 0$;

6: **while** $\forall j \in \mathcal{J} : \sum_k \sum_n \mu_{jnk}^{(t_1)} - \mu_{jnk}^{(t_1-1)} \neq 0$ **do**

7:     Each node j: assign labels $\mu_{jnk}^{(t_1)}$ based on (6.2).

8:     $\mathbf{z}_k^{(0)} = 0$, select $\alpha^{(t_2)}$; $t_2 = 0$;

9:     **while** $\|\mathbf{m}_k^{(t_2)} - \mathbf{m}_k^{(t_2-1)}\| \leq \epsilon_{glo}$ **do**

10:       Each node j: send $\mathbf{g}_{jk}^{(t_2)}$ to server; $\mathbf{z}_k^{(t_2+1)}$ by (6.4a);

11:       Server computes $\mathbf{z}_k^{(t_2+1)}$ by (6.4a), $\mathbf{m}_k^{(t_2+1)}$ by (6.4b) and send it back to nodes;

12:     **end while**

13:     $t_1 = t_1 + 1$;

14: **end while**

6.4   Simulations

In the following figures, the algorithms are labeled as *UFLBin* and *UFLSOM*. The centralized k-means algorithm is used as the performance benchmark, with all using the initialization and perturbation techniques described in Sect. 6.2. In addition, uniform gradient weighting, labeled as *UFLUni* is used for comparison to show the benefit of proper weighting.

Data are generated at random from $K = 16$ classes, with vectors from each class generated from a symmetric Gaussian distributions with means $\mathbf{m}_1 = [-7.5, -7.5]^T$, $\mathbf{m}_2 = [-7.5, -2.5]^T$, $\mathbf{m}_3 = [-7.5, 2.5]^T$, $\mathbf{m}_4 = [-7.5, 7.5]^T$, $\mathbf{m}_5 = [-2.5, -7.5]^T$, $\mathbf{m}_6 = [-2.5, -2.5]^T$, $\mathbf{m}_7 = [-2.5, 2.5]^T$, $\mathbf{m}_8 = [-2.5, 7.5]^T$, $\mathbf{m}_9 = [2.5, -7.5]^T$, $\mathbf{m}_{10} = [2.5, -2.5]^T$, $\mathbf{m}_{11} = [2.5, 2.5]^T$, $\mathbf{m}_{12} = [2.5, 7.5]^T$, $\mathbf{m}_{13} = [7.5, -7.5]^T$, $\mathbf{m}_{14} = [7.5, -2.5]^T$, $\mathbf{m}_{15} = [7.5, 2.5]^T$, $\mathbf{m}_{16} = [7.5, 7.5]^T$ similar to Fig. 6.2. Each cluster contains 50 points. The network consists of $J = 5$ sensing agents that are placed randomly and observe data according to energy detector for Rayleigh fading with a probability $P(d_j, \mathrm{SNR}_j) = 1 - \exp(-\frac{\alpha \mathrm{SNR}_j}{d_j})$. d_j is the distance between the jth agent and the data points, and SNR_j is the signal-to-noise ratio in linear scale at the jth agent. α is an observation parameter and is set to 0.025 in the simulations.

Convergence results are shown in Fig. 6.4. It is clear that the proposed algorithms take relatively the same number of global iterations (indexed by t_1 in Algorithm 11) to converge. It can be observed that the convergence rate is similar to that of a centralized k-means algorithm. Clustering performance vs. SNR is shown in Fig. 6.5, where four metrics are used, namely, objective value, variation of

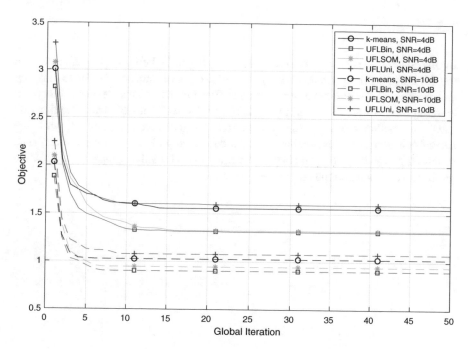

Fig. 6.4 Convergence of the proposed algorithms

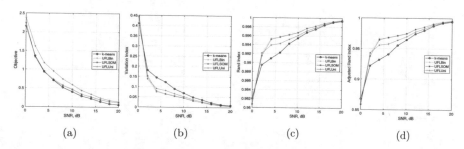

Fig. 6.5 Performance of proposed algorithms. (**a**) Performance of objective value. (**b**) Performance of VI. (**c**) Performance of RI. (**d**) Performance of ARI

information (VI), Rand Index (RI) and Adjusted Rand Index (ARI). First metric reflects accuracy of centroid estimation, while the other three describes correctness of data labels. Both UFLBin and UFLSOM perform better than UFLUni and close to k-means in terms of objective. Hence, the proposed algorithms are able to achieve superior performance, while the SOM based method requires less signaling than the bin method. Overall, the proposed methods have been shown to achieve good results compared to uniform gradient weighting and the centralized k-means.

6.5   Summary

In this chapter, we consider unsupervised learning tasks being implemented within the federated learning framework to satisfy stringent requirements for lowlatency and privacy of the emerging applications. Two DA based unsupervised federated learning schemes are discussed to tackle the problem of non-IID data. The first one uses a bin method that suffers from the curse of dimensionality problem, while the second can overcome this problem, albeit incurring slightly more computations. The proposed methods have been shown to achieve good results compared to uniform gradient weighting and the centralized k-means.

Part III
Federated Learning Applications in Wireless Networks

Chapter 7
Wireless Virtual Reality

Abstract This chapter introduce the use of federated learning (FL) for wireless virtual reality (VR) applications. In particular, we first explain why we use to use FL for wireless VR applications. Then, we provide a detailed literature review of using FL for VR applications. We then introduce a representative work that focuses on the use of FL for the analysis and predictions of orientation and mobility of VR users so as to reduce break in presences of VR users.

7.1 Motivation

Deploying virtual reality (VR) applications over wireless networks is an essential stepping stone towards flexible deployment of pervasive VR applications [3, 169, 170]. However, to enable a seamless and immersive wireless VR experience, it is necessary to introduce novel wireless networking solutions that can meet stringent quality-of-service (QoS) requirements of VR applications [171, 172]. In wireless VR, any sudden drops in the data rate or increase in the delay can negatively impact the users' VR experience (e.g., due to interruptions in VR video streams). Due to such an interruption in the virtual world, VR users will experience *breaks in presence (BIP)* events that can be detrimental to their immersive VR experience. While the fifth-generation (5G) new radio supports operation at high frequency bands as well as flexible frame structure to minimize latency, the performance of communication links at high frequencies is highly prone to blockage. That is, if an object blocks the wireless link between the BS and a VR user, the data rate can drop significantly and lead to a BIP. In addition to wireless factors such as delay and data rate, behavioral metrics related to each VR user such as the user's *awareness* can also induce BIP. Awareness is defined as each wireless VR user's perceptions and actions in its individual VR environment. Therefore, to minimize the BIP of VR users, it is necessary to jointly consider all of the wireless environment and user-specific metrics that cause BIP, such as link blockage, user location, user orientation, user association, and user awareness.

7.2   Existing Works

Recently, several works have studied a number of problems related to wireless VR networks [173–182]. The work in[173] developed a multipath cooperative route scheme to enable VR wireless transmissions. In [174], the authors develop a framework for mobile VR delivery by leveraging the caching and computing capabilities of mobile VR devices. The authors in [175] study the problem of supporting visual and haptic perceptions over wireless cellular networks. A communications-constrained mobile edge computing framework is proposed in [176, 177] to reduce wireless resource consumption. The work in [178] proposes a concrete measure for the delay perception of VR users. The authors in [179] present a scheme of proactive computing and high-frequency, millimeter wave (mmWave) [183] transmission for wireless VR networks. In [180], the authors design several experiments for quantifying the performance of tile-based 360° video streaming over a real cellular network. The works in [181, 182] studied the problem of 360° content transmission. However, most of these existing works do not provide a comprehensive BIP model that accounts for the transmission delay, the quality of VR videos, VR application type, and user awareness. Moreover, the prior art in [173–182] does not jointly consider the impact of the users' body movements when using mmWave communications.

To address this challenge, machine learning techniques can be used to predict the users' movements and proactively determine the user associations that can minimize BIP. However, in prior works on machine learning for user movement predictions [184–189], the data for each user's movement must be collected by its associated BS. However, in real mobile VR scenarios, users will move and change their association and the data related to their movement is dispersed across multiple BSs. In such scenarios, the BSs may not be able to continuously share collected user data among each other, due to the high overhead of data transmission. Moreover, sending all the information to a centralized processing server will cause very large delays that cannot be tolerated by VR applications. Thus, centralized machine learning algorithms such as in [185–189] will not be useful to predict real-time movements of the VR users. To this end, *a distributed learning framework that can be trained by the collected data at each BS is needed.*

Recently, a number of existing works such as in [67, 72, 74, 75] studied important problems related to the implementation of distributed learning over wireless networks. While interesting, these prior works [67, 72, 74, 75] that focus on the optimization of the performance of distributed learning algorithms such as federated learning do not consider the use of distributed learning to optimize the performance of wireless networks. In particular, these existing works [67, 72, 74, 75] do not consider the use of distributed learning algorithms to predict users' orientations and locations to reduce the BIP of wireless VR users.

7.3 Representative Work

Next, we introduce a novel framework for minimizing BIP within VR applications that operate over wireless networks [73]. In particular, we first introduce a mathematical model of a new BIP metric that jointly considers VR application type, the delay of VR video and tracking information transmission, VR video quality, and the users' awareness. To minimize the BIP of wireless VR users, we introduce a federated ESN [66, 190] learning algorithm that enables BSs to locally train their machine learning algorithms using the data collected from the users' locations and orientations. Then, the BSs can cooperatively build a learning model by sharing their trained models to predict the users' locations and orientations. Based on these predictions, we perform fundamental analysis to find an efficient user association for each VR user that minimizes the BIP.

7.3.1 System Model

Consider a cellular network that consists of a set $\mathcal{B}$ of B BSs that service a set $\mathcal{U}$ of U VR users. In this model, BSs act as VR controllers that can collect the tracking information related to the users' movements via VR sensors and use the collected data to generate the VR videos for their associated users, as shown in Fig. 7.1. In particular, the uplink is used to transmit tracking information such as users' locations and orientations from the VR devices to the BSs, while the downlink is used to transmit VR videos from BSs to VR users. For user association, the VR users can associate with different BSs for uplink and downlink data transmissions. Different from prior works such as in [171, 174–176, 178–181] that assume the VR users to be static, we consider a practical scenario in which the locations and orientations of the VR users will impact the VR application performance.

Fig. 7.1 The architecture of a wireless VR network. In this architecture, the Sub-6 GHz uplink is used to transmit tracking information and the mmWave downlink is used to transmit VR videos

Transmission Model

We consider both uplink and downlink transmission links between BSs and VR users. The VR users can operate at both mmWave and sub-6 GHz frequencies [191–193]. The VR videos are transmitted from BSs to VR users over the 28 GHz band. Meanwhile, the tracking information is transmitted from VR devices to their associated BSs over a sub-6 GHz frequency band. This is due to the fact that sub-6 GHz frequencies with limited bandwidth cannot support the large data rates required for VR video transmissions. However, it can provide reliable communications for sending small data sized users' tracking information.

Let (x_{it}, y_{it}) be the Cartesian coordinates for the location of user i at time t and S be the data size of each user's tracking information, including location and orientation. S depends on the VR system (i.e., HTC Vive [194] or Oculus [195]). The data rate for transmitting the tracking information from VR user i to BS j is given by:

$$c_{ij}^{UL}(x_{it}, y_{it}) = F^{UL}\log_2\left(1 + \frac{P_u g_{ij} d_{ij}^{-\beta}(x_{it}, y_{it})}{\sum\limits_{k \in \mathcal{U}_i} P_u g_{kj} d_{kj}^{-\beta}(x_{kt}, y_{kt}) + \rho^2}\right), \quad (7.1)$$

where F^{UL} is the bandwidth of each subcarrier, U_j^{UL} is the number of VR users associated with BS j over uplink, $\mathcal{U}_i$ is the set of VR users that use the same subcarriers with user i, P_u is the transmit power of each VR user (assumed equal for all users), g_{ij} is the Rayleigh channel gain, d_{ij} is the distance between VR user i and BS j, and ρ^2 is the noise power.

In the downlink, antenna arrays are deployed at BSs to perform directional beamforming over the mmWave frequency band. For simplicity, a sectored antenna model [196] is used to approximate the actual array beam patterns. This simplified antenna model consists of four parameters: the half-power beamwidth ϕ, the boresight direction θ, the antenna gain of the mainlobe Q, and the antenna gain of the sidelobe q. Let φ_{ij} be the phase from BS j to VR user i. The antenna gain of the transmission link from BS j to user i is:

$$G_{ij} = \begin{cases} Q, & \text{if } |\varphi_{ij} - \theta_j| \leqslant \frac{\phi}{2}, \\ q, & \text{if } |\varphi_{ij} - \theta_j| > \frac{\phi}{2}. \end{cases} \quad (7.2)$$

Since the VR device is located in front of the VR user's head, the mmWave link will be blocked, if the user rotates. Let χ_{it} be the orientation of user i at time t and ϑ be the maximum angle using which BS j can directly transmit VR videos to a user without any human body blockage. ϕ'_{ij} denotes the phase from user i to BS j.

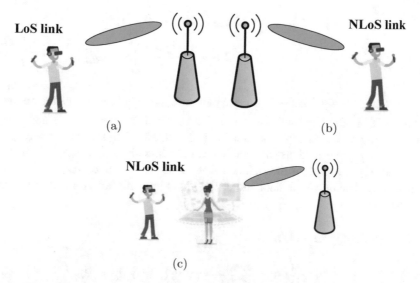

Fig. 7.2 VR video transmission over LoS/NLoS links. (**a**) LoS links. (**b**) NLoS links caused by the user's own body. (**c**) NLoS links caused by other user's body

For user i, the blockage effect, $b_i(\chi_{it})$, caused by its own body can be given by:

$$b_i(\chi_{it}) = \begin{cases} 1, & \text{if } \left| \varphi'_{ij} - \chi_{it} \right| \leqslant \vartheta, \\ 0, & \text{if } \left| \varphi'_{ij} - \chi_{it} \right| > \vartheta. \end{cases} \tag{7.3}$$

We assume that each VR user's body constitutes a single blockage area and n_{ijt} represents the number of VR users located between user i and BS j at time t. If there are no users located between user i and BS j that block the mmWave link, i.e., $(b_i(\chi_{it}) + n_{ij} = 0$, then, as shown in Fig. 7.2a, the communication link between user i and BS j is line-of-sight (LoS). If the mmWave link between user i and BS j is blocked by the user i's own body (as shown in Fig. 7.2b, $b_i(\chi_{it}) = 1$) or blocked by other users located between user i and BS j (as shown in Fig. 7.2c, $+n_{ij} > 0$), then the communication link between user i and BS j is said to be non-line-of-sight (NLoS). From (7.3), we can see that $b_i(\chi_{it})$ and n_{ij} can be directly determined by the users' orientations and locations.

Considering path loss and shadowing effects, the path loss for a LoS link and a NLoS link between VR user i and BS j in dB will be given by Semiari et al. [196]:

$$h_{ij}^{\text{LoS}}(x_{it}, y_{it}) = 10\varpi_{\text{LoS}} \log \left(d_{ij}(x_{it}, y_{it}) \right)$$
$$+ 20 \log \left(\frac{d^0 f_c 4\pi}{v} \right) + \mu_{\sigma_{\text{LoS}}}, \tag{7.4}$$

$$h_{ij}^{\text{NLoS}}(x_{it}, y_{it}) = 10\varpi_{\text{NLoS}} \log\left(d_{ij}(x_{it}, y_{it})\right)$$
$$+ 20\log\left(\frac{d^0 f_c 4\pi}{\nu}\right) + \mu_{\sigma_{\text{NLoS}}}, \tag{7.5}$$

where $20\log\left(\frac{d_0 f_c 4\pi}{\nu}\right)$ is the free space path loss. Here, d^0 represents the reference distance, f_c is the carrier frequency and ν is the light speed. ϖ_{LoS} and ϖ_{NLoS} represent the path loss exponents for the LoS and NLoS links, respectively. $\mu_{\sigma_{\text{LoS}}}$ and $\mu_{\sigma_{\text{NLoS}}}$ represent Gaussian random variables with zero mean, respectively. σ_{LoS} and σ_{NLoS} represent the standard deviations for LoS and NLoS links in dB, respectively. The downlink data rate of VR video transmission from BS j to user i is given by:

$$c_{ij}^{\text{DL}}\left(x_{it}, y_{it}, b_i(\chi_{it}), n_{ij}\right)$$
$$= \begin{cases} F^{\text{DL}}\log_2\left(1 + \frac{P_{\text{B}}G_{ij}}{10^{h_{ij}^{\text{LoS}}/10}\rho^2}\right), \text{if } b_i(\chi_{it}) + n_{ij} = 0, \\ \\ F^{\text{DL}}\log_2\left(1 + \frac{P_{\text{B}}G_{ij}}{10^{h_{ij}^{\text{NLoS}}/10}\rho^2}\right), \text{if } b_i(\chi_{it}) + n_{ij} > 0, \end{cases} \tag{7.6}$$

where F^{DL} is the bandwidth allocated to each user and P_{B} is the transmit power of each BS j which is assumed to be equal for all BSs. Since the downlink uses mmWave links, we assume that, due to directivity, interference in (7.6) can be neglected, as done in [197].

Break in Presence Model

In a VR application, the notion of a BIP represents an event that leads the VR users to realize that they are in a fictitious, virtual environment, thus ruining their immersive experience. In other words, a BIP event transitions a user from the immersive virtual world to the real world [198]. For wired VR, BIP can be caused by various factors such as hitting the walls/ceiling, loss of tracking with the device, tripping on wire cords, or talking to another person from the real world [198]. For wireless VR, BIP can be also caused by the delay of VR video and tracking information transmission, the quality of the VR videos received by the VR users, and inaccurate tracking information received by BSs.

To model such BIP, we jointly consider the delay of VR video and tracking information transmission and the quality of the VR videos. We first define a vector $l_{i,t}\left(c_{ij}^{\text{DL}}(x_{it}, y_{it}, b_i(\chi_{it}), n_{ij})\right) = [l_{i1,t}, \ldots, l_{iN_L,t}]$ that represents a VR video that user i received at time t with $l_{ik,t} \in \{0, 1\}$. $l_{ik,t} = 0$ indicates that pixel k is not successfully received by user i, and $l_{ik,t} = 1$, otherwise. We also define a vector $m_{i,t}(G_A) = [m_{i1,t}, \ldots, m_{iN_L,t}]^{\text{T}}$ that represents the weight of the importance of each pixel constructing a VR video, where $m_{ik,t} \in [0, 1]$ and G_A represents a VR application such as an immersive VR game or a VR video. $m_{ik,t} = 1$ indicates that pixel k is one of the most important elements for the generation of G_A. The value of $m_{ik,t}$ depends on the compression used for the VR video. In each VR application G_A, a number of pixels can be compressed at the BS and recovered by the user and, hence, these pixels are not important. However, the pixels that cannot be compressed by the BS are important and must be transmitted to the VR users. Therefore, each pixel will have different importance and $m_{ik,t} \in [0, 1]$. Then, the BIP of VR user i caused by the wireless transmission will be given by:

$$\omega_{it}\left(x_{it}, y_{it}, \chi_{it}, \boldsymbol{a}_{i,t}^{\text{UL}}, \boldsymbol{a}_{i,t}^{\text{DL}}\right) =$$

$$\mathbb{1}\left\{ \frac{A}{a_{ij,t}^{\text{UL}} c_{ij}^{\text{UL}}(x_{it}, y_{it})} + \frac{D(l_{i,t})}{a_{ik,t}^{\text{DL}} c_{ik}^{\text{DL}}} > \gamma_D \vee l_{i,t} m_{i,t}(G_A) < \gamma_Q \right\}. \tag{7.7}$$

where c_{ik}^{DL} is short for $c_{ik}^{\text{DL}}(x_{it}, y_{it}, b_i(\chi_{it}), n_{ik})$. Meanwhile, $l_{i,t}$ is simplified for $l_{i,t}\left(a_{ik,t}^{\text{DL}} c_{ik}^{\text{DL}}(x_{it}, y_{it}, b_i(\chi_{it}), n_{ik})\right)$, respectively, $\mathbb{1}_{\{x\}} = 1$ if x is true, and otherwise, we have $\mathbb{1}_{\{x\}} = 0$. $\mathbb{1}_{\{x\}} \vee \mathbb{1}_{\{y\}} = 1$ as y or x is true, $\mathbb{1}_{\{x\}} \vee \mathbb{1}_{\{y\}} = 0$, otherwise. $\boldsymbol{a}_{i,t}^{\text{UL}} = \left[a_{i1,t}^{\text{UL}}, \dots, a_{iB,t}^{\text{UL}} \right]$ is a vector that represents user i's uplink association with $a_{ik,t}^{\text{UL}} \in \{0, 1\}$ and $\sum_{k\in\mathcal{B}} a_{ik,t}^{\text{UL}} = 1$. Similarly, $\boldsymbol{a}_{i,t}^{\text{DL}} = \left[a_{i1,t}^{\text{DL}}, \dots, a_{iB,t}^{\text{DL}} \right]$ is a vector that represents user i's downlink association with $a_{ik,t}^{\text{DL}} \in \{0, 1\}$ and $\sum_{k\in\mathcal{B}} a_{ik,t}^{\text{DL}} = 1$. γ_D and γ_Q represent the target delay and video quality requirements, respectively. In (7.7), A represents the data size of the tracking information, $\frac{A}{c_{ij}^{\text{UL}}(x_{it}, y_{it})}$ represents the time used for tracking information transmission from user i to BS j, $D\left(l_{i,t}\left(c_{ik}^{\text{DL}}(x_{it}, y_{it}, b_i(\chi_{it}), n_{ik})\right)\right)$ represents the data size of a VR video, and $\frac{D(l_{i,t}(c_{ik}^{\text{DL}}(x_{it}, y_{it}, b_i(\chi_{it}), n_{ik})))}{c_{ik}^{\text{DL}}(x_{it}, y_{it}, b_i(\chi_{it}), n_{ik})}$ represents the transmission latency for sending the tracking information from BS k to user i. For simplicity, hereinafter, ω_{it} is referred as $\omega_{it}\left(x_{it}, y_{it}, \chi_{it}, \boldsymbol{a}_{i,t}^{\text{UL}}, \boldsymbol{a}_{i,t}^{\text{DL}}\right)$. (7.7) shows that if the delay of VR video and tracking information transmission exceeds the target delay threshold allowed by VR systems or the quality of the VR video cannot meet the video requirement, users will experience a BIP ($\omega_{it} = 1$). From (7.7), we can also see that, the BIP of user i caused by wireless transmission depends on user i's location, orientation, VR applications, and user association. (7.7) captures the BIP caused by wireless networking factors such as transmission delay and video quality. Next, we define a BIP model that jointly considers wireless transmission, the VR application type, and the users' awareness. The BIP of user i can be given by Chung et al. [199]:

$$P_i\left(x_{it}, y_{it}, G_A, \chi_{it}, a_{i,t}^{\text{UL}}, a_{i,t}^{\text{DL}}\right) = \frac{1}{T}\sum_{t=1}^{T}\left(G_A + \omega_{it} + G_A\omega_{it} + \epsilon_i + \epsilon_{G_A|i} + \epsilon_B\right),$$

$$(7.8)$$

where ϵ_i is user i's awareness, $\epsilon_{G_A|i}$ is the joint effect caused by user i's awareness and VR application G_A, and ϵ_B is a random effect. ϵ_i, $\epsilon_{G_A|i}$, and ϵ_B follow a Gaussian distribution [199] with zero mean and variances σ_i^2, $\sigma_{G_A|i}^2$, and σ_B^2, respectively. In (7.8), the value of $P_i\left(x_{it}, y_{it}, G_A, \chi_{it}, a_{i,t}^{\text{UL}}, a_{i,t}^{\text{DL}}\right)$ quantifies the average number of BIP that user i can identify during a period. From (7.8), we can see that, as the VR application for user i changes, the BIP value will change. For example, a given user watching VR videos will experience fewer BIP compared to a user engaged in an immersive first-person shooting game. This is due to the fact that in an immersive game environment, users are fully engaged with the virtual environment, as opposed to some VR applications that require the user to only watch VR videos. In (7.8), we can also see that the BIP depend on the users' awareness. This means that different users will have different actions and perceptions when they interact with the virtual environment and, hence, different VR users may experience different levels of BIP.

Problem Formulation

From (7.8), we can see that the BIP of each user depends on this user's location, orientation, and selected BSs. By using an effective learning algorithm to predict the users' locations and orientations, the BSs can proactively determine the users' association to improve the downlink and uplink data rates and minimize BIP for each VR user. The BIP minimization problem is:

$$\min_{a_{i,t}^{\text{UL}}, a_{i,t}^{\text{DL}}} \sum_{i\in\mathcal{U}} P_i\left(\hat{x}_{it}, \hat{y}_{it}, G_A, \hat{\chi}_{it}, a_{i,t}^{\text{UL}}, a_{i,t}^{\text{DL}}\right) \qquad (7.9)$$

$$\text{s. t.} \quad U_j \leqslant V, \quad \forall j \in \mathcal{B}, \qquad (7.9a)$$

$$a_{ij,t}^{\text{UL}}, a_{ij,t}^{\text{DL}} \in \{0, 1\}, \quad \forall i \in \mathcal{U}, \forall j \in \mathcal{B}, \qquad (7.9b)$$

$$\sum_{j\in\mathcal{B}} a_{ij,t}^{\text{UL}} = 1, \ \sum_{j\in\mathcal{B}} a_{ij,t}^{\text{DL}} = 1, \quad \forall i \in \mathcal{U}, \qquad (7.9c)$$

where $\hat{x}_{it}$, $\hat{y}_{it}$, and $\hat{\chi}_{it}$ are the predicted locations and orientation of user i at time t, which depend on the actual historical locations and orientation of user i. U_j is the number of VR users associated with BS j over downlink and V is the maximum number of users that can be associated with each BS. (7.9b) and (7.9c) show that each user can associate with only one uplink BS and one downlink BS. From (7.9), we can see that the BIP of each user will depend on the user association as well as the users' locations and orientations. Meanwhile, the user association depends on the locations and orientations of the VR users. If the BSs perform the user association without knowledge of the locations and orientations of the users, the body blockage between the user-BS transmission links can potentially be significant, thus increasing the BIP of each user. Therefore, the BSs must use historical information related to the users' locations and orientations to determine the user association. As the users' locations and orientations will continuously change as time elapses, BSs must proactively determine the user association to reduce the BIP of VR users. In consequence, it is necessary to introduce a machine learning algorithm to predict the users' locations and orientations in order to determine the user association and minimize BIP of VR users. In the previous defined model, the user association changes as the users' location and orientation vary with time. Consequently, each BS that connects to a given VR user can only collect partial information about this user's locations and orientation. However, a BS cannot rely on partial information to predict each user's location and orientation. Moreover, since a given VR user will change its association, the data pertaining to this VR user's movement will be located at multiple BSs. Hence, traditional centralized learning algorithms that are implemented by a given BS cannot predict the entire VR user's locations and orientations without knowing the user's data collected by other BSs. To overcome the challenges mentioned previously, we introduce a distributed federated learning framework that can predict the location and orientation of each VR user as the training data related to each user's locations and orientations is located at multiple BSs.

7.3.2   Federated Echo State Learning for Predictions of the Users' Location and Orientation

Federated learning is a decentralized learning algorithm that can operate by using training datasets that are distributed across multiple devices (e.g., BSs). For our system, one key advantage of federated learning is that it can allow multiple BSs to locally train their local learning model using their collected data and cooperatively build a learning model by sharing their locally trained models. Compared to existing federated learning algorithms [54] that use matrices to record the users' behavior and cannot analyze the correlation of the users' behavior data, we propose an ESN-based federated learning algorithm that can use an ESN to efficiently analyze the data related to the users' location and orientation. The ESN based FL algorithm enables the BSs to collaboratively generate a global ESN model to predict the whole set of locations and orientations for each user without transmitting the collected data to other BSs. However, if the BSs use the centralized learning algorithms for the orientation and location predictions, the BSs must use the data collected from all BSs to train the algorithm. ESNs have two unique advantages: simple training process and the ability to analyze time-dependent data [66]. Since the data that is related to the orientation and locations of the users is time-dependent and the users' orientation and locations will change frequently, we must use ESNs that can efficiently analyze time-dependent data and converge quickly to obtain the prediction results on time and determine the user association. Next, we first introduce the components of the federated ESN learning model. Then, we explain the entire procedure of using our federated ESN learning algorithm to predict the users' locations and orientation.

Components of Federated ESN Learning Algorithm

A federated ESN learning algorithm consists of five components: (a) agents, (b) input, (c) output, and (d) local ESN model, which are specified as follows:

- *Agent*: In our system, we need to define an individual federated ESN learning algorithm to predict the location and orientation of each VR user. Meanwhile, each user's individual federated ESN learning algorithm must be implemented by all BSs that have been associated with this user. Each BS j must implement U learning algorithms to predict the locations and orientations of all users.
- *Input:* The input of the federated ESN learning algorithm that is implemented by BS j for the predictions of each VR user i is defined by a vector $v_{ij} = \left[v_{ij,1}, \cdots, v_{ij,T}\right]^{\mathrm{T}}$ that represents the information related to user i's location and orientation where $v_{ij,t} = \left[\xi_{ij1,t}, \ldots, \xi_{ijN_x,t}\right]$ represents user i's information related to location and orientation at time t. This information includes user i's locations, orientations, and VR applications. N_x is the number of properties that constitute a vector $v_{ij,t}$. The input of the proposed algorithm will be combined with the ESN model to predict users' orientation and locations. BSs will use these predictions to determine user associations.
- *Output:* For each user i, the output of the federated ESN learning algorithm at BS j is a vector $y_{ij,t} = \left[\hat{y}_{ijt+1}, \ldots, \hat{y}_{ijt+Y}\right]$ of user i's locations and orientations where $\hat{y}_{ijt+k} = \left[\hat{x}_{it+k}, \hat{y}_{it+k}, \hat{\chi}_{it+k}\right]$ with $\hat{x}_{it+k}$ and $\hat{y}_{it+k}$ being the predicted location coordinates of user i at time $t + k$ and $\hat{\chi}_{it+k}$ being the estimated orientation of user i at $t + k$. Y is the number of future time slots that a federated ESN learning algorithm can predict. The predictions of the locations and orientations can be used to determine the user's association.
- *Local ESN model:* For each BS j, a local ESN model is used to build the relationship between the input of all BSs and the predictions of the users' location and orientation, as shown in Fig. 7.4. The local ESN model consists of the input weight matrix $W_j^{\mathrm{in}} \in \mathbb{R}^{N_W \times T}$, recurrent matrix $W_j \in \mathbb{R}^{N_W \times N_W}$, and the output weight matrix $W_j^{\mathrm{out}} \in \mathbb{R}^{Y \times (N_W + T)}$. The values of W_j^{in} and W_j are generated randomly. However, the output weight matrix W_j^{out} need to be trained according to the inputs of all BSs.

Fig. 7.3 Architectures of deep ESN models. (**a**) A series ESN model. (**b**) A parallel ESN model

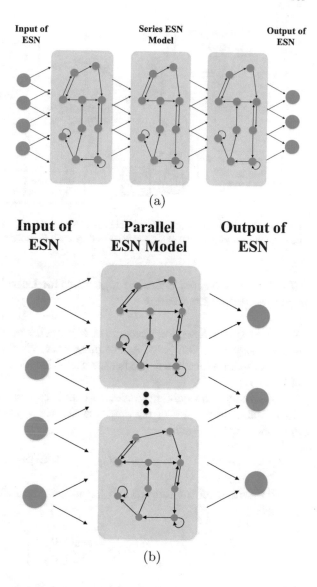

(a)

(b)

We introduce three ESN models: *single ESN model, series ESN model, and parallel ESN model*. In the single ESN model, an ESN is directly connected to the input and output. Moreover, as shown in Fig. 7.3, series and parallel ESN models connect single ESN models in series and parallel, respectively. Each ESN model has its own advantage for our problem. In particular, a single ESN model can converge faster than a series ESN model and a parallel ESN model. A parallel ESN model has a larger memory capacity than a series ESN model. A series ESN model can decrease the prediction errors in the training process.

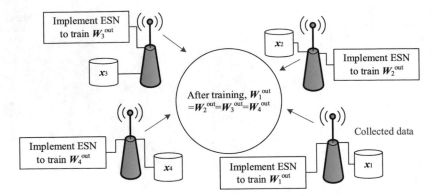

Fig. 7.4   The implementation of the ESN based federated learning. Here, the data is located at the BSs and the learning model W_j^{out} that is trained by each BS's collected data is the local model

ESN Based Federated Learning Algorithm for Users' Location and Orientation Predictions

Next, we explain the entire procedure of training the proposed ESN-based federated learning algorithm. Our purpose of training ESN is to find an optimal output weight matrix in order to accurately predict the users' locations and orientations, as shown in Fig. 7.4.

To introduce the training process, we first explain the state of the neurons in ESN. The neuron states of the proposed algorithm implemented by BS j for the predictions of user i are:

$$\boldsymbol{\mu}_{j,t} = \boldsymbol{W}_j \boldsymbol{\mu}_{j,t-1} + \boldsymbol{W}_j^{\text{in}} \boldsymbol{v}_{ij,t}. \tag{7.10}$$

Based on the states of neurons and the inputs, the ESN can estimate the output, which is:

$$\hat{\boldsymbol{y}}_{ij,t} = \boldsymbol{W}_{j,t}^{\text{out}} \begin{bmatrix} \boldsymbol{v}_{ij,t} \\ \boldsymbol{\mu}_{j,t} \end{bmatrix}. \tag{7.11}$$

From (7.11), we can see that, in order to enable an ESN to predict the users' locations and orientations, we only need to adjust the value of the output weight matrix. However, each BS can collect only partial data for each user and, hence, we need to use a distributed learning algorithm to train the ESNs. To introduce the distributed learning algorithm, we first define two matrices which are given by:

$$\boldsymbol{H}_j = \begin{bmatrix} \boldsymbol{v}_{ij,1} \ \boldsymbol{\mu}_{j,1} \\ \vdots \\ \boldsymbol{v}_{ij,T} \ \boldsymbol{\mu}_{j,T} \end{bmatrix} \text{ and } \boldsymbol{E}_j = \begin{bmatrix} \boldsymbol{e}_{ij,1}, \ldots, \boldsymbol{e}_{ij,T} \end{bmatrix}, \tag{7.12}$$

where $e_{ij,t}$ is the desired locations and orientations of each VR user, given the ESN input $v_{ij,t}$. Then, the training purpose can be given as follows:

$$\min_{W^{\text{out}}} \frac{1}{2} \left(\sum_{j=1}^{B} \left\| W^{\text{out}} H_j^{\text{T}} - E_j \right\|^2 \right) + \frac{\lambda}{2} \left\| W^{\text{out}} \right\|. \tag{7.13}$$

(7.13) is used to find the optimal global output weight matrix W^{out} according to which the BSs can predict the entire users' locations and orientations without the knowledge of the users' data collected by other BSs. From (7.13), we can see that, each BS j needs to adjust its output weight matrix W_j^{out} and find the optimal output weight matrix W^{out}. After the learning step, we have $W_j^{\text{out}} = W^{\text{out}}$, which means that when the learning algorithm converges, the local model of each BS will converge to the global model.. A standard update policy of W_j^{out} for the augmented Lagrangian problem in (7.13) is given by Scardapane et al. [200]:

$$W_{j,t+1}^{\text{out}} = \varsigma^{-1} \left[I - H_j^{\text{T}} \left(\varsigma I + H_j H_j^{\text{T}} \right) H_j^{\text{T}} \right] \\ \times \left(H_j^{\text{T}} E_j - n_{j,t} + \varsigma W_t^{\text{out}} \right), \tag{7.14}$$

where ς is the learning rate and W_t^{out} is the optimal output weight matrix that the ESN model of each BS needs to find. From (7.14), we can see that $W_{j,t+1}^{\text{out}}$ is the output weight matrix that is generated at BS j. $W_{j,t+1}^{\text{out}}$ can only be used to predict partial locations and orientations given the users' data collected by BS j. $W_{j,t+1}^{\text{out}}$ is different from the output weight matrices of other BSs. The optimal output weight matrix is given by:

$$W_{t+1}^{\text{out}} = \frac{B \varsigma \hat{W}_{t+1}^{\text{out}} + B \hat{n}_t}{\lambda + \varsigma B}, \tag{7.15}$$

where $\hat{W}_{t+1}^{\text{out}}$ and $\hat{n}_{t+1}^{\text{out}}$ can be calculated as follows:

$$\hat{W}_{t+1}^{\text{out}} = \frac{1}{B} \sum_{j=1}^{B} W_{j,t+1}^{\text{out}}, \quad \hat{n}_t = \frac{1}{B} \sum_{j=1}^{B} n_{j,t}. \tag{7.16}$$

From (7.14) to (7.16), we can see that the global output weight matrix W^{out} is based on (7.15) and (7.16) while the local output weight matrix W_j^{out} is based on (7.14). In (7.14), $n_{j,t}$ is the deviation between the output weight matrix $W_{j,t+1}^{\text{out}}$ of each BS j and the optimal output weight matrix W_{t+1}^{out} that the ESN model of each BS needs to converge, which is given by:

$$n_{j,t+1} = n_{j,t} + \gamma \left(W_{j,t+1}^{\text{out}} - W_{t+1}^{\text{out}} \right). \tag{7.17}$$

W_{t+1}^{out} is the global optimal output weight matrix that can be used to predict the entire locations and orientations of a given user. This means that using W_{t+1}^{out}, each BS can predict the entire user's locations and orientations as the BS only collects partial data related to the user's locations and orientations. As time elapses, $W_{j,t+1}^{\text{out}}$ will finally converge to W_{t+1}^{out}. In consequence, all of BSs can predict the entire locations and orientations of each user. To measure the convergence, we define two vectors which can be given by $r_{j,t} = W_{j,t}^{\text{out}} - W_t^{\text{out}}$ and $s_{j,t} = W_t^{\text{out}} - W_{t-1}^{\text{out}}$. As $\|r_{j,t+1}\| \leqslant \gamma_A$ or $\|s_{j,t}\| \leqslant \gamma_A$, the proposed algorithm converges. γ_A is determined by the BSs. Since the minimization function in (7.13) is a convex function, the BSs are guaranteed to find an optimal output weight matrix that satisfy $\|r_{j,t+1}\| \leqslant \gamma_A$ or $\|s_{j,t}\| \leqslant \gamma_A$. As γ_A increases, the accuracy of the predictions and the number of iterations decrease. Therefore, BSs need to jointly account for the time used for training ESN and the prediction accuracy to determine the value of γ_A. In fact, the ESN As the learning algorithm converges, each BS can use its own ESN to predict the entire location and orientation of each VR user. According to these predictions, BSs can determine the user association to minimize the BIP of VR users.

7.3.3 Memory Capacity Analysis

To improve the prediction accuracy of the proposed algorithm, we analyze the memory capacity of the proposed ESN model. The memory capacity quantifies the ability of each ESN to record the historical locations and orientations of each VR user. As the memory capacity of the ESNs increases, the ESNs can record more historical data related to users' locations and use this information to achieve better prediction[1] for the users' locations and orientations. The analysis of the ESN memory capacity will be used for the choice of the ESN models for the predictions of the users' locations and orientations. Next, we derive closed-form expressions of the memory capacity of the three ESN models that we described in Sect. 7.3.2, namely, the single ESN model, the parallel ESN model, and the series ESN model. Note that, our previous work [201] analyzed the memory capacity for a centralized parallel ESN model. In contrast, here, we analyze the memory capacity for three ESN models used for federated learning.

---

[1] Here, as the size of the recorded data increases, the ESNs can use more historical data to build a relationship between historical orientations and locations, and future orientations and locations. Hence, the ESN prediction accuracy improves.

We assume that the input of each ESN model at time t is m_t and the output of each ESN model is z_t. Then, the memory capacity of each ESN model is given by:

$$M = \sum_{k=1}^{\infty} \frac{\text{Cov}^2(m_{t-k}, z_t)}{\text{Var}(m_t)\text{Var}(z_t)}, \tag{7.18}$$

where Cov and Var represent the covariance and variance operators, respectively. In (7.18), $\frac{\text{Cov}^2(m_{t-k}, z_t)}{\text{Var}(m_{t-k})\text{Var}(z_t)}$ captures the correlation between the ESN input m_{t-k} at time $t - k$ and the ESN output z_t at time t. $\frac{\text{Cov}^2(m_{t-k}, z_t)}{\text{Var}(m_{t-k})\text{Var}(z_t)} = 1$ indicates that m_{t-k} and z_t are related which means that the output z_t includes the information of m_{t-k} and, hence, the ESN can record input m_{t-k} at time t. $\frac{\text{Cov}^2(m_{t-k}, z_t)}{\text{Var}(m_t)\text{Var}(z_t)} = 0$ indicates that m_{t-k} and z_t are unrelated, which means that z_t does not include any information related to m_{t-k} and, hence, the ESN cannot record m_{t-k}. In consequence, M represents the total number of historical input data that each ESN can record. The recurrent matrix W in each ESN model is given by:

$$W_l = \begin{bmatrix} 0 & 0 & \cdots & w \\ w & 0 & 0 & 0 \\ 0 & \ddots & 0 & 0 \\ 0 & 0 & w & 0 \end{bmatrix}, \tag{7.19}$$

and the input weight matrix is given by $W^{\text{in}} = \left[w_1^{\text{in}}, \ldots, w_{N_W}^{\text{in}} \right]^{\text{T}}$. We also define a matrix that will be used to derive the memory capacity of the ESNs, which can be given by:

$$V = \begin{bmatrix} w_1^{\text{in}} & w_{N_W}^{\text{in}} & \cdots & w_2^{\text{in}} \\ w_2^{\text{in}} & w_1^{\text{in}} & \cdots & w_3^{\text{in}} \\ \vdots & \vdots & \cdots & \vdots \\ w_{N_W}^{\text{in}} & w_{N_W-1}^{\text{in}} & \cdots & w_1^{\text{in}} \end{bmatrix}. \tag{7.20}$$

Based on the above definitions, we can invoke our result from [190, Theorem 2] to derive the memory capacity of single ESN model, which can be given as follows.

Corollary 7.1 (Single ESN Model) *Given the recurrent matrix W and the input matrix W^{in} that guarantees the matrix V regular, the memory capacity of the single ESN model is:*

$$M = N_W - 1 + w^{2N_W}. \tag{7.21}$$

Proof Given the input stream vector $m_{\ldots t} = [m_1, \ldots, m_{t-1}, m_t]$, we can calculate the activations $\mu_{j,t}$ using (7.10). The output weight matrix of the ESN model can be given by $W^{\text{out}} = R^{-1}p_k$, where $R = \mathbb{E}\left[\mu_t\left(\mu_t\right)^{\text{T}}\right]$ represents the covariance matrix with $\mu_t = \left[\mu_{1,t}, \ldots, \mu_{N_W,t}\right]$ and $p_k = \mathbb{E}\left[\mu_t m_{t-k}\right]$. Assume that $w^{\text{in}}_{N_W \ldots 1} = \left[w^{\text{in}}_{N_W}, w^{\text{in}}_{N_W-1}, \ldots, w^{\text{in}}_1\right]$ and $\text{rot}_k\left(w^{\text{in}}_{N_W \ldots 1}\right)$ is an operator that rotates vector $w^{\text{in}}_{N_W \ldots 1}$ by k positions to the right. We have $W^{\text{out}} = (1 - w^{2N_W})w^k A^{-1}\text{rot}_k(w^{\text{in}}_{1 \ldots N_W})$, where $A = V^{\text{T}}\Gamma^2 V$ with $\Gamma = \left(1, w, \ldots, w^{N_W-1}\right)$. Based on W^{out}, we can the covariance of the output with the k-slot delayed input, which is given by $\text{Cov}(z_t, m_{t-k}) = (1 - w^{2N_W})w^{2k}\sigma^2\zeta_k$. We can also obtain $\text{Var}(z_t) = \mathbb{E}[z_t z_t] = (1 - w^{2N_W})w^{2k}\sigma^2\zeta_k$. Since $\text{Var}(m_t) = \sigma^2$, we have $M = \sum_{k=1}^{\infty} \frac{\text{Cov}^2(m_{t-k}, z_t)}{\text{Var}(m_t)\text{Var}(z_t)} = N_W - 1 + w^{2N_W}$.

From Corollary 7.1, we can see that the memory capacity of the single ESN model depends on the number of neurons and values of the recurrent matrix. Corollary 7.1 also shows that the memory capacity of the single ESN model will not exceed N_W. That means the single ESN model based federated learning algorithm can only record N_W locations or orientations.

Next, we derive the memory capacity of the parallel ESN model, which can be given by the following theorem.

Theorem 7.1 (Parallel ESN) *Given a parallel ESN model during which L ESN models are parallel connected with each other, each ESN model's input weight matrix W^{in} that guarantees the matrix V regular and recurrent matrix W, then the memory capacity of each parallel ESN can be given by:*

$$M = N_W - 1 + w^{2N_W}. \tag{7.22}$$

Proof See in [73]

Theorem 7.1 shows that the memory capacity of a parallel ESN model is similar to the memory capacity of a single ESN. Hence, adding multiple ESN models will not increase the memory capacity. This is due to the fact that, in a parallel ESN model, there is no connection among the ESNs, as shown in Fig. 7.3b. Therefore, the input of the parallel ESN model will separately connect to each single ESN and, hence, the parallel ESN models do not need to use more neurons to record the input data compared to the single ESN model. Theorem 7.1 also shows that the memory capacity of a parallel ESN depends on the number of neurons in each ESN model and the values of the recurrent weight matrix of each ESN model. Accordingly, we can increase the value of output weight matrix and the number of neurons in each ESN model to increase the memory capacity of the parallel ESN models. As the memory capacity of the parallel ESN models increases, BSs can record more users' data to predict the users' locations and orientations accurately. Next, we derive the memory capacity of the series ESN model.

Theorem 7.2 (Series ESN Model) *Given a series ESN model during which L ESN models are series connected with each other, a recurrent matrix W of each ESN* *model, and each ESN model's input weight matrix $\boldsymbol{W}^{\text{in}}$ that guarantees the matrix $\boldsymbol{V}$ regular, the memory capacity of each series ESN model is:*

$$M = \left(1 - w^{2N_W}\right)^{L-1} \left(N_W - 1 + w^{2N_W}\right). \tag{7.23}$$

Proof See in [73].

From Theorem 7.2, we can see that the memory capacity of each series ESN model is smaller than the memory capacity of a single ESN or a series ESN. Theorem 7.2 also shows that the memory capacity of each series ESN model decreases as the number of ESN models L increases. Thus, it would be better to use a single ESN model or a parallel ESN model to predict the users' locations and orientations.

Theorems 7.1 and 7.2 derive the memory capacities of the parallel ESN model and the series ESN model with single input. Next, we formulate the memory capacity of a single ESN model given multiple inputs, which is given by the following theorem.

Theorem 7.3 (Multi-Input Single ESN) *Consider a single ESN with a recurrent matrix $\boldsymbol{W}$, input vector $\boldsymbol{m}_t = [m_{1t}, \ldots, m_{Kt}]$, the input weight matrix $\boldsymbol{W}^{\text{in}}$ that guarantees the matrix $\boldsymbol{V}$ regular, the memory capacity of each single ESN is*

$$M = \left(\frac{\sum_{l=1}^{K} \sigma_l^2}{\sum_{k=1}^{K} \sum_{n=1}^{K} \rho_{kn} \sigma_k \sigma_n}\right)^2 \left(N_W - 1 + w^{2N_W}\right), \tag{7.24}$$

where ρ_{kn} represents the correlation coefficient between input m_{kt} and m_{nt}.

Proof See in [73].

From Theorem 7.3, we can observe that the correlation among input elements in vector $\boldsymbol{m}_t$ will affect the memory capacity of each ESN model. In particular, as the correlation of the input data increases, the memory capacity of the ESN model increases. This is because the ESN can use more input data to predict the users' locations and orientations, hence improving the predictions accuracy. Therefore, it would be better to jointly predict the users' locations and orientations.

Theorems 7.1–7.3 allow each BS to determine its ESN model, the number of neurons N_W in each ESN model, and the values of the recurrent matrix $\boldsymbol{W}$ as the size of the data collected by each BS changes. A parallel ESN model has a larger memory capacity compared with the series ESN model and is more stable than the single ESN model, and, hence, a parallel ESN model can record more historical data to predict the users' orientations and locations so as to improve the prediction accuracy. As the prediction accuracy is improved, the BSs can determine the user association more accurately. Hence, the BIP of the users can be minimized. Therefore, we use the parallel ESN model in our proposed algorithm.

7.3.4   User Association for VR Users

Based on the analysis presented in Sects. 7.3.2 and 7.3.3, each BS can predict the users' locations and orientations. Next, we explain how to use these predictions to find the user association for each VR user. Given the predictions of the locations and orientations, the BIP minimization problem in (7.9) can be rewritten as follows:

$$\min_{a_{i,t}^{\text{UL}},a_{i,t}^{\text{DL}}} \sum_{i\in\mathcal{U}} P_i \left(\hat{x}_{it}, \hat{y}_{it}, G_A, \hat{\chi}_{it}, a_{i,t}^{\text{UL}}, a_{i,t}^{\text{DL}} \right). \tag{7.25}$$

We use the reinforcement learning algorithm given in [202] to find a sub-optimal solution of the problem in (7.25). In the reinforcement learning algorithm given in [202], the actions are the user association schemes, the states are the strategies of other BSs, and the output is the estimated BIP. Hence, this reinforcement learning algorithm can learn the VR users state and exploit different actions to adapt the user association according to the predictions of the users' locations and orientations. After the learning step, each BS will find a sub-optimal user association to service the VR users. To simplify the learning process and improve the convergence speed, we first select the uplink user association scheme. This is because as the uplink user association is determined, the BSs that the users can associate in downlink will be determined, as follows:

Proposition 7.1 *Given the predicted location and orientation of user i at time t as well as the uplink user association $a_{i*,t}^{\text{UL}}$, the downlink cell association for a VR user i is:*

$$a_{ik,t}^{\text{DL}} = \mathbb{1}\left\{ \frac{D\left(l_{i,t}\left(a_{ik,t}^{\text{DL}},c_{ik}^{\text{DL}}\right)\right)}{a_{ik,t}^{\text{DL}}c_{ik}^{\text{DL}}} \leqslant \gamma_{\text{D}} - \frac{A}{a_{i*,t}^{\text{UL}}c_{i*}^{\text{UL}}(\hat{x}_{it},\hat{y}_{it})} \right\}$$

$$\wedge \mathbb{1}\left\{ l_{i,t}\left(a_{ik,t}^{\text{DL}},c_{ik}^{\text{DL}}\right)m_{i,t}(G_A) \geqslant \gamma_{\text{Q}} \right\}, \tag{7.26}$$

where $c_{ik}^{\text{DL}}\left(\hat{x}_{it}, \hat{y}_{it}, b_i\left(\hat{\chi}_{it}\right), n_{ik}\right)$ is short for c_{ik}^{DL} and $a_{ik,t}^{\text{DL}}$ is the downlink user association obtained in (7.26). $c_{i*}^{\text{UL}}(x_{it},y_{it})$ is the uplink data rate of user i.

Proof For downlink user association, each VR user i needs to find a BS that can guarantee the transmission delay and VR video quality. Since we have determined the user association over uplink, the maximum time used for VR video transmission can be given by $\gamma_{\text{D}} - \frac{A}{a_{i*,t}^{\text{UL}}c_{i*}^{\text{UL}}(\hat{x}_{it},\hat{y}_{it})}$. Consequently, user i needs to connect with a BS that can satisfy the transmission delay requirement of user i, i.e., $\frac{D\left(l_{i,t}\left(a_{ik,t}^{\text{DL}}c_{ik}^{\text{DL}}(\hat{x}_{it},\hat{y}_{it},b_i(\hat{\chi}_{it}),n_{ik})\right)\right)}{a_{ik,t}^{\text{DL}}c_{ik}^{\text{DL}}(\hat{x}_{it},\hat{y}_{it},b_i(\hat{\chi}_{it}),n_{ik})} \leqslant \gamma_{\text{D}} - \frac{A}{a_{i*,t}^{\text{UL}}c_{i*}^{\text{UL}}(\hat{x}_{it},\hat{y}_{it})}$. Moreover, user i needs to associate with a BS that can meet the requirement of VR video quality, i.e., $l_{i,t}\left(a_{ik,t}^{\text{DL}}c_{ik}^{\text{DL}}\left(\hat{x}_{it}, \hat{y}_{it}, b_i\left(\hat{\chi}_{it}\right), n_{ik}\right)\right)m_{i,t}(G_A) \geqslant \gamma_{\text{Q}}$. Thus, if BS k can satisfy the conditions: $\dfrac{D\left(l_{i,t}\left(a_{ik,t}^{\mathrm{DL}}c_{ik}^{\mathrm{DL}}(\hat{x}_{it},\hat{y}_{it},b_i(\hat{\chi}_{it}),n_{ik})\right)\right)}{a_{ik,t}^{\mathrm{DL}}c_{ik}^{\mathrm{DL}}(\hat{x}_{it},\hat{y}_{it},b_i(\hat{\chi}_{it}),n_{ik})} \leqslant \gamma_{\mathrm{D}} - \dfrac{A}{a_{i*,t}^{\mathrm{UL}}c_{i*}^{\mathrm{UL}}(\hat{x}_{it},\hat{y}_{it})}$ and

$l_{i,t}\left(a_{ik,t}^{\mathrm{DL}}c_{ik}^{\mathrm{DL}}(\hat{x}_{it},\hat{y}_{it},b_i(\hat{\chi}_{it}),n_{ik})\right)m_{i,t}(G_A) \geqslant \gamma_{\mathrm{Q}}$, user i can associate with it. This completes the proof.

From Proposition 7.1, we can see that the user association of each user i depends on user i's location and orientation. Proposition 7.1 shows that, for each user i, the uplink user association will affect the downlink user association. This is due to the fact that the VR system has determined the total transmission delay of each user. As a result, when the uplink user association is determined, the uplink transmission delay and the requirement of the downlink transmission delay will be determined.

7.3.5 Simulation Results and Analysis

For our simulations, we consider a circular area with radius $r = 500$ m, $U = 20$ wireless VR users, and $B = 5$ BSs distributed uniformly. To simulate blockage, each user is considered as a two-dimensional point. For simplicity, we ignore the altitudes of the BSs and the height of the users. If blockage points are located between a user and its BS, the communication link will be considered to be NLoS. Real data traces for locations are collected from 50 students at the Beijing University of Posts and Telecommunications. The locations of each student is collected every hour during 9:00 a.m.–9:00 p.m. For orientation data collection, we searched 25 videos related to a first-person shooter game from youTube. Then, we input these VR videos to HTC Vive devices. The HTC Vive developer system can directly measure the movement of the VR videos using HTC Vive devices. We arbitrarily combine one user's locations with one orientation for each VR user. In simulations, a parallel ESN model is used for the proposed algorithm due to its stability and large memory capacity. The other system parameters are listed in Table 7.1. For comparison purposes, we consider the deep learning algorithm in [187] and the ESN algorithm in [188], as two baseline schemes. The deep learning algorithm in [187] is a deep autoencoder that consists of multiple layers of restricted Boltznann machines. The centralized ESN-based learning algorithm in [188] is essentially a single layer ESN algorithm. The input and output of the centralized ESN and deep learning algorithms are similar to the proposed algorithm. However, for the deep learning algorithm and the centralized ESN algorithm, each BS can use only its collected data to train the learning model. Both the centralized and deep learning algorithms are trained in an offline manner. All statistical results are averaged over a large number of independent runs.

Figures 7.5 and 7.6 show the predictions of the VR users' locations and orientations as time elapses. To simplify the model training, the collected data related to locations and orientations are mapped to $[-0.5, 0.5]$. The orientation and location of each user are, respectively, mapped by the function $\frac{\chi_{it}}{360°} - 0.5$ and

Table 7.1 System
parameters

Parameters	Values	Parameters	Values
P_B	30 dBm	d_0	5 m
P_U	10 dBm	f_c	28 GHz
σ	−94 dBm	c	3×10^8 m/s
F^{UL}	10 Mbit	$\varpi_{LoS}, \varpi_{NLoS}$	2, 2.4
F^{DL}	10 Mbit	$\mu_{\sigma_{LoS}}, \mu_{\sigma_{NLoS}}$	5.3, 5.27
N_W	30	G_A	11
β	2	γ_D	10 ms
M	15 dB	γ_Q	0.8
m	0.7 dB	$\sigma_i{}^2$	0.193
ϕ	30°	A	50 kbits
Y	10	T	5
γ	0.5	λ	0.005
w	0.98	L	3
V	10	ϑ	2
$\sigma^2_{G_A|i}$	0.151	$\sigma_B{}^2$	0.05

$\frac{z}{z_{max}} - 0.5$ where $z = \left(\sum_{n=1}^{\hat{x}_{it}+\hat{y}_{it}} n \right) \times \hat{y}_{it}$. From Figs. 7.5 and 7.6, we observe that the
proposed algorithm can predict the users' locations and orientations more accurately
than the centralized ESN and deep learning algorithms. Figure 7.6b and c also show
that the prediction error mainly occurs at time slot 8–12. This is due to the fact that
the proposed algorithm can build a learning model that predicts the entire locations
and orientations of each user. In particular, the output weight matrices of all ESN
algorithms implemented by each BS will converge to a common matrix. Hence, BSs
can predict the entire locations and orientations of each VR user.

Figure 7.7 shows how the total BIP of all VR users changes as the number of BSs
varies. From Fig. 7.7, we can see that, as the number of BSs increases, the total BIP
of all VR users decreases. That is because as the number of BSs increases, the VR
users have more connection options. Hence, the blockage caused by human bodies
will be less severe, thereby improving the data rates of VR users. Figure 7.7a also
shows that the proposed algorithm can achieve up to 16% and 26% reduction in
the number of BIP, respectively, compared to centralized ESN algorithm and deep
learning algorithm for a network with 9 BSs. These gains stem from the fact that
the centralized ESN and deep learning algorithms can partially predict the locations
and orientation of each VR user as they rely only on the local data collected by
a BS. In contrast, the proposed algorithm facilitates cooperation among BSs to
build a learning model that can predict the entire users' locations and orientations.
Figure 7.7b shows that the proposed algorithm using a parallel ESN model can
achieve up to 8% and 14% gains in terms of the total BIP of all users compared to
the proposed algorithm with a single ESN model and with a series model. Clearly,
compared to a single ESN, using a parallel ESN model can increase the stability
of the proposed algorithm. Meanwhile, the memory capacity of a parallel model is

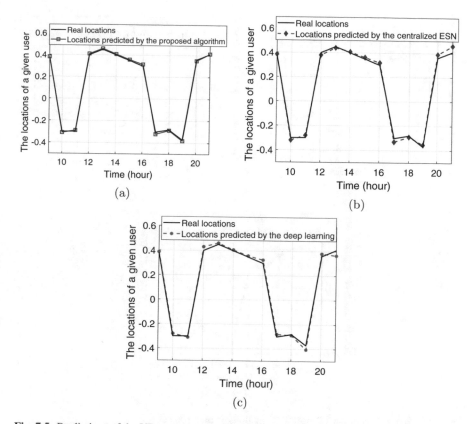

Fig. 7.5 Predictions of the VR users' locations as time elapses

larger than a series ESN model thus improving the prediction accuracy and reducing BIP for users.

In Fig. 7.8, we show how the total BIP of all VR users changes with the number of VR users. This figure shows that, with more VR users, the total BIP of all VR users increases rapidly due to an increase in the uplink delay, as the sub-6 GHz bandwidth is shared by more users. Figure 7.8 also shows that the gap between the proposed algorithm and the centralized ESN algorithm decreases as more VR users are present in the network.. Clearly, with more VR users, it becomes more probable that a user located between a given VR user and its associated BS blocks the mmWave link. Thus, as the number of users increases, more VR users will receive their VR videos over NLoS links and, the total BIP significantly increases.

In Fig. 7.9, we show the CDF for the VR users' BIP for all three algorithms. Figure 7.9 shows that the BIP of almost 98% of users resulting from the considered algorithms will be larger than 10. This is due to the fact that the BIP will also be caused by other factors such as VR applications and user's awareness. In Fig. 7.9, we can also see that the proposed algorithm improves the CDF of up to 38% and 71% gains at a BIP of 25 compared to the centralized ESN and deep learning algorithms,

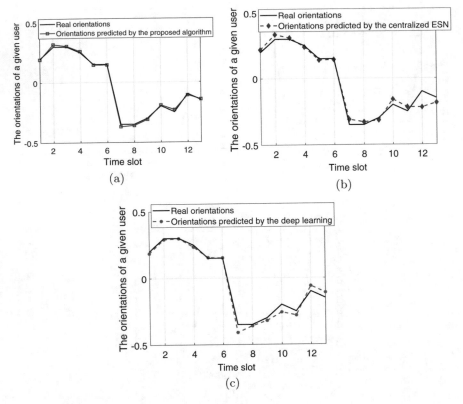

Fig. 7.6 Predictions of the VR users' orientations as time elapses

respectively. These gains stem from the fact the ESNs are effective at analyzing the time related location and orientation data and, hence, they can accurately predict the users' locations and orientations.

7.4   Summary

In this chapter, we have introduced the use of FL for wireless VR applications. In particular, we have first introduced the advantages of using FL for wireless VR applications. Then, we have provided a detailed literature review of the use of machine learning for wireless VR applications. We then have introduced a representative work that focuses on the development of a novel framework for minimizing BIP within VR applications that operate over wireless networks. To this end, we have developed a BIP model that jointly considers the VR applications, transmission delay, VR video quality, and the user's awareness. We have then formulated an optimization problem that seeks to minimize the BIP of VR users

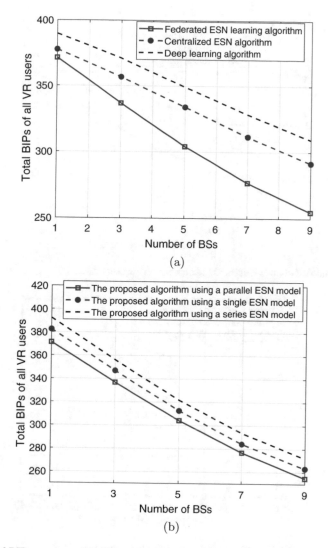

Fig. 7.7 Total BIP experienced by VR users as the number of BSs varies

by predicting users' locations and orientations, as well as determining the user association. To solve this problem, we have developed a novel federated learning algorithm based on echo state networks. The proposed federated ESN algorithm enables the BSs to train their ESN with their locally collected data and share these models to build a global learning model that can predict the entire locations and orientations of each VR user. To improve the prediction accuracy of the proposed algorithm, we derive a closed-form expression of the memory capacity for ESNs to determine the number of neurons in each ESN model and the values of the recurrent

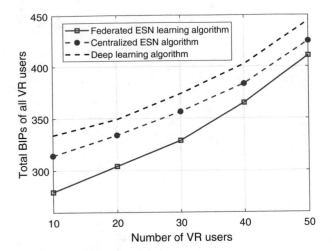

Fig. 7.8 Total BIP of all VR users as the number of VR users varies

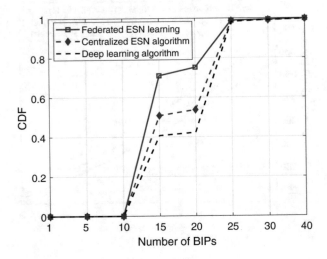

Fig. 7.9 CDFs of the BIP resulting from the different algorithms

weight matrix. Using these predictions, each BS can determine the user association in both uplink and downlink. Simulation results have shown that, when compared to the centralized ESN and deep learning algorithms, the proposed approach achieves significant performance gains of BIP.

Chapter 8
Vehicular Networks and Autonomous Driving Cars

Abstract In this chapter, we discuss the role of federated learning for vehicular networks. Due to the high mobility of autonomous cars, there might not be seamless connectivity of the end-devices within cars with the roadside units, and thus traditional federated learning might not work well. To overcome this challenge, we introduced a dispersed federated learning framework for autonomous driving cars. We formulate a dispersed federated learning cost optimization problem and proposed an iterative scheme. Finally, we present extensive simulation results to validate the proposal.

8.1 Introduction and State of Art

Federated learning has been proposed by Google as a distributed machine learning paradigm to push the computation of artificial intelligence (AI) applications into more and more end devices while protecting the privacy of end users [142]. In federated learning, a central server sends an initialized global deep neural network (DNN) model to clients as the first step. Based on the initialized global DNN model, clients separately train local DNN models with their local data as the second step. Instead of directly sending their local data, clients send the trained local DNN models back to the central server as the third step. The above steps are repeated in multiple rounds until the training accuracy of the global DNN model meets the requirement of the central server. Due to the above advantages, federated learning has been applied to many application scenarios, such as financial applications [142], virtual keyboard applications [203], Internet of Things [141], and electronic health applications [204].

Vehicular edge computing (VEC) is a fast-developing vehicular technology, where vehicles and roadside servers at the network edge contribute communication, computation, storage and data resources to close proximity of vehicular users [13, 205–207]. With the rapid penetration of intelligent connected vehicles (ICV), there is an urgent need to study federated learning in VEC as an important technical framework to meet the ever-increasing demands of AI applications in vehicular networks. In the following section, we consider image classification as a typical AI

© The Author(s), under exclusive license to Springer Nature Singapore Pte Ltd. 2021   179
C. S. Hong et al., *Federated Learning for Wireless Networks*, Wireless Networks,
https://doi.org/10.1007/978-981-16-4963-9_8

application in VEC [208]. As we know, the images captured from on-board cameras usually contain sensitive information with individual privacy of the vehicular clients. Using federated learning in VEC is beneficial in exploiting vehicular images for DNN training while protecting their privacy. For example, the vehicular clients use on-board cameras to capture images, which are classified and labeled by automatic labeling technology [209]. After that, the vehicular clients are selected by the central server to participate in federated learning in a supervised fashion and generate global and local DNN model updates.

The major challenge of federated learning in VEC is two folds. On the one hand, the diversity of image quality may cause severe loss of the accuracy of model aggregation. In VEC, the captured images generally suffer from motion blur, noise, and distortions [210], especially motion blur that is usually with different levels for different vehicular clients. During local training, the local DNNs are tuned according to the local images, and therefore, only work with the best accuracy under the specific statistics of the motion blur. As a result, the overall accuracy of the aggregated global DNN model will severely degrade if inappropriate local DNN models are involved. On the other hand, the diversity of computation capability has an impact on the efficiency of model aggregation. The difference in computation capability leads to different latency of training local DNN models. For synchronization, the central server performs model aggregation only after receiving all the local DNN models. This means that the vehicular clients with low computation capability hinder the efficiency of model aggregation [16].

To improve the performance of model updates, the authors in [211] design a greedy algorithm to find out as many clients with high computation capability and good wireless channel condition as possible. Under bandwidth and time limitation, the authors in [212] design a heuristic algorithm to assign the clients who are willing to upload their local data to a central server. The uploaded data is constructed for approximately independent and identically distributed (i.i.d.), which increases the classification accuracy. In these studies, it is not practical to assume that the clients contribute their resources without the compensation of the cost of consuming resources. Accordingly, the authors in [26, 213, 214] utilize game theory to attract clients to share their resources. In [213], the authors use the Stackelberg game to incentive clients to contribute their data resources for improving the learning accuracy of the model. Similarly, the authors in [214] use the Stackelberg game to incentive clients to contribute their computation resources for reducing the latency of model training. In [26], the authors adopt the Stackelberg game to study the interaction between participating clients and an edge server. The interaction includes the strategies of participating clients and the edge server, i.e., local relative accuracy and reward. The participating clients make optimal local relative accuracy to maximize their benefits. Then the edge server makes optimal reward to its benefit, which improves the global accuracy of model training. But the above studies assume that the central server is aware of clients' data quality, computation capability, energy state, and willingness to participate, namely information asymmetry. To overcome the information asymmetry, contract theory is a powerful tool to model the incentive mechanism [215, 216]. The authors in

[28] use a multi-weight subjective logic model to design a reputation-based worker selection scheme for reliable federated learning. Then, they use contract theory to stimulate high-reputation workers with high-quality data to participate in model training, which reduces the latency of model training. In addition, a consortium blockchain is used to manage the reputation in a decentralized manner. The above existing work focus on mobile edge computing (MEC) [211, 217] and distributed networks [28, 212–214]. In this chapter, we discuss federated learning in VEC, which is important for generalizing AI applications in ICV, although it has not been reported in other work. Table 8.1 gives the comparison of existing related work.

8.2 Vehicular Networks

As shown in Fig. 8.1, the general framework of federated learning in VEC consists of the following components:

- **Central Server:** Central server plays a core role in the procedure of federated learning. It communicates with vehicular clients to collect the updated local DNN models and perform model aggregation. We take image classification as a typical AI application in VEC. DNN-based image classification has been widely used in autopilot and interactive navigation for ICV, as well as object tracking and event detection in ITS [218, 219]. To obtain high accuracy and efficiency of model aggregation, the central server should evaluate the image quality and computation capability of vehicular clients, and select the "fine" models from vehicular clients.
- **Vehicular Client:** Vehicular clients are equipped with a set of built-in sensors, such as cameras, GPS, tachographs, lateral acceleration sensors, and also accommodate storage space, computation and communication resources [219]. The built-in sensors are used to capture images that may be preprocessed for data augment. After that, the preprocessed images are classified and labeled by automatic labeling technology [209], and are cached in vehicular clients. After receiving a request from a central server, vehicular clients separately train local DNN models with their local images. Vehicular clients send updated the local DNN models to the central server for model aggregation.

Based on the principle of federated learning, the original algorithm, i.e., federated averaging (FedAvg), will randomly assign some vehicular clients to perform tasks of training the local DNN models [18]. The selected vehicular clients have diverse image quality and computation capability, which reduces the accuracy and efficiency of model aggregation. To cope with the above dilemma, we propose a selective model aggregation approach.

Table 8.1 A comparison about client selection for federated learning in edge computing and distributed networks

Ref.	Network type	Client heterogeneity	Information feature	Approach
[211]	Mobile edge computing	Computation capability and communication condition	Information symmetry	Greedy algorithm for selecting clients with high computation capability and good wireless channel condition to improve model performance
[212]	Distributed network	Computation capability and willingness to upload data	Information symmetry	Heuristic algorithm for constructing i.i.d. data to improve model performance
[213]	Distributed network	Amount of data	Information symmetry	Stackelberg game for improving learning accuracy of model
[214]	Distributed network	Computation capability	Information symmetry	Stackelberg game for reducing latency of model training
[26]	Mobile edge computing	Computation capability and communication condition	Information symmetry	Stackelberg game for improving global accuracy of model training
[28]	Distributed network	Computation capability	Information asymmetry	Contract based incentive mechanism for reducing latency of model training

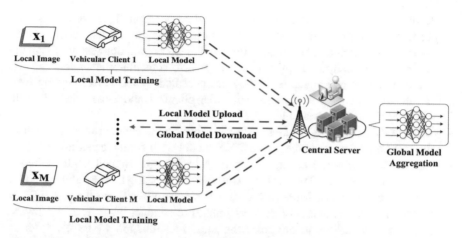

Fig. 8.1 A general framework of federated learning in vehicular edge computing

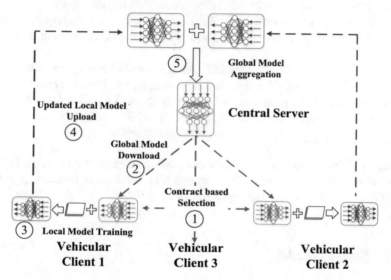

Fig. 8.2 Selective model aggregation. Three vehicular clients are illustrated in the case, where vehicular clients 1 and 2 are finally selected while vehicular client 3 is not, according to the contract based procedure

8.2.1   Selective Model Aggregation

As shown in Fig. 8.2, the main procedure of selective model aggregation has the following steps.

- **Step 1: Contract based selection:** The central server initializes a global DNN model denoted as $\mathbf{w}(0)$. Based on the historical records of vehicular clients, the central server evaluates their image quality and computation capability. The details about the utilized evaluation method are presented in Sect. 8.2.2. The central server designs two-dimensional contract items for vehicular clients. Each item includes the amount of images, the amount of computation resources and the reward. All the contract items are broadcasted to vehicular clients periodically. The contract items are signed if they are accepted by the corresponding type of vehicular clients. For example, vehicular clients 1 and 2 are selected while vehicular client 3 is not, in Fig. 8.2.

- **Step 2: Global model download:** After confirming the contract items, vehicular clients 1 and 2 download the global DNN model $\mathbf{w}(0)$ from the central server.
- **Step 3: Local model training:** According to the predesigned contract items, vehicular clients 1 and 2 train a local DNN model by using their local images and computation resources. More specifically, vehicular client 1 uses the global DNN model $\mathbf{w}(0)$ and a number of x_1 local images to conduct the forward-backward propagation algorithm to minimize the local loss function $F_1(\mathbf{w}(0))$. After E rounds of local iterations, vehicular client 1 updates the local DNN model $\mathbf{w}_1^E(0)$. Similarly, vehicular client 2 updates the local DNN model $\mathbf{w}_2^E(0)$.
- **Step 4: Updated local model upload:** To meet synchronization requirements, the updated local DNN models $\mathbf{w}_1^E(0)$ and $\mathbf{w}_2^E(0)$ are sent to the central server in time.
- **Step 5: Global model aggregation:** After receiving the updated local DNN models $\mathbf{w}_1^E(0)$ and $\mathbf{w}_2^E(0)$, the central server aggregates them to update the global DNN model, which generates the global DNN model $\mathbf{w}(1)$. Also, the central server aggregates the local loss functions $F_1(\mathbf{w}_1^E(0))$ and $F_2(\mathbf{w}_2^E(0))$ as a new global loss function $F(\mathbf{w}(1)) = \frac{x_1 F_1(\mathbf{w}_1^E(0)) + x_2 F_2(\mathbf{w}_2^E(0))}{x_1 + x_2}$ [18].

Steps 1–5 form one global iteration (i.e., one communication round). In the k-th global iteration, the change of the global loss function is denoted as $\Delta F_k = F(\mathbf{w}(k)) - F(\mathbf{w}(k-1))$, namely global loss decay [220].

8.2.2 System Model

We now consider a general scenario where a central server schedules a set of vehicular clients (denoted as M). In the model aggregation, the heterogeneity of resources among the vehicular clients affects the accuracy and efficiency of model aggregation. In other words, the diverse image quality and computation capability affect the accuracy and efficiency of model aggregation, respectively. Each vehicular client knows exactly its image quality and computation capability, but the image quality and computation capability are not available to the central server. This means that there exists asymmetric information between the vehicular clients and the central server. To overcome the above problem, the central server can leverage contract theory to design an incentive mechanism to motivate the vehicular clients to participate in the model aggregation. In contract theory, an employer makes optimal contracts for the employees when the employer does not know the privacy information of each employee [215]. Here contract theory is used to model the interactions between the central server and the vehicular clients under information asymmetry. The central server acts as the employer and offers different contract items to the vehicular clients. The vehicular clients act as the employees and select the contract items matching their own types.

Next, we define the image quality and computation capability of vehicular clients. Based on the image quality and computation capability, we define the utilities of vehicular clients and the types of vehicular clients. Finally, we model the utility of central server.

Image Quality

Due to the mobility of vehicles, the images captured by on-board cameras generally suffer from motion blur, noise, and distortion [221, 222]. The noise and distortion in different vehicular clients may follow identical statistical distribution, while the motion blur level varies with instantaneous velocity of each vehicular client [210]. For depicting the motion blur level caused by instantaneous velocity, we utilize a geometric model to illustrate the relationship between an object of interest and the on-board camera. According to the model, the motion blur level can be implicitly predicted by observing the instantaneous velocity of each vehicular client. By Cortés-Osorio et al. [210], we have

$$v' = \frac{\sigma l}{H[s\cos(\delta) - (g+l)\sin(\delta)]}, \tag{8.1}$$

where v' is the relative velocity between velocity v of vehicular client and velocity v_o of the object, σ is the perpendicular distance from the pinhole to the starting point of an object, l is the length of the motion blur on the image plane, H is the exposure time interval, s is the camera focal length, δ is the angle between the image plane and the motion direction, and g is the starting position of the object on the image plane. We denote the charge-coupled device (CCD) pixel size in the horizontal direction as Q, and have

$$L = \frac{v'H[s\cos(\delta) - QG\sin(\delta)]}{v'HQ\sin(\delta) + \sigma Q}, \tag{8.2}$$

where G and L are the starting position of the object and the level of motion blur in the image (in pixels), respectively. As shown in Fig. 8.3, considering the case where the image plane and the motion direction are parallel ($\delta = 0$), and the object of interest is static ($v_o = 0$), Eq. (8.2) is transformed into

$$L = \frac{vsH}{\sigma Q}, \tag{8.3}$$

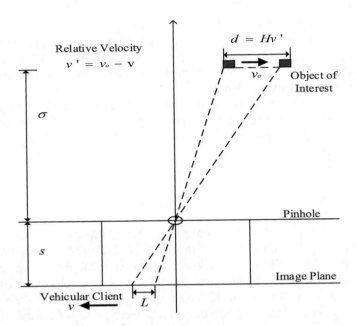

Fig. 8.3 Geometric model for image quality analysis

where $\frac{sH}{\sigma Q}$ is a parameter of the on-board camera. The equation directly shows that low instantaneous velocity means the low motion blur level.

Based on the motion blur level, we try to evaluate the image quality. By Pei et al. [208], we consider that when the motion blur level of training images is more similar to that of testing images, the higher the classifying accuracy is resulted. As a consequence, we measure the image quality by function β that has the form as

$$\beta = \beta(L, L_t),\tag{8.4}$$

where L_t is the given motion blur level of testing images. Function β has the following characteristics. If L is approximated to L_t, $\beta(L, L_t)$ is larger; and vice versa. If $|L_1 - L_t| = |L_2 - L_t|$ and $L_1 < L_2$, we have $\beta(L_1, L_t) \geq \beta(L_2, L_t)$. To satisfy the above characteristics, $\beta(L, L_t)$ is defined by

$$\beta(L, L_t) = \begin{cases} e^{q_1(L-L_t)}, 0 \leq L \leq L_t, \\ e^{-q_2(L-L_t)}, L_t \leq L, \end{cases}\tag{8.5}$$

where q_1 and q_2 are two predefined constants. In Fig. 8.4, we shows an example of function β, where $q_1 = 0.5$, $q_2 = 0.8$ and $L_t = 6$.

In the k-th global iteration, based on the image quality, we express the valuation function of vehicular client m as

$$r_{k,m} = \beta_{k,m} h_k(p_{k,m}),\tag{8.6}$$

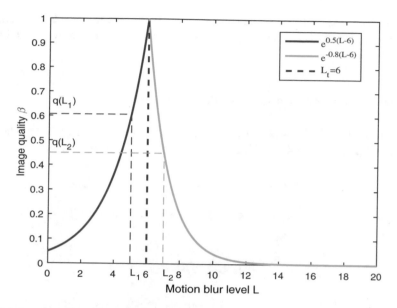

Fig. 8.4 Image quality with motion blur level

where $\beta_{k,m}$ is the image quality for vehicular client m, $p_{k,m}$ is the reward for contributing images and computation resources to the central server, and $h_k(p_{k,m})$ is a revenue function which is increased with the increasing of the reward $p_{k,m}$. The similar valuation function appears in [223].

Computation Capability

For vehicular client m, contributing images and computation resources incurs a cost of resource consumption, which is denoted as

$$c_{k,m} = \alpha_{k,m} x_{k,m} + E_{k,m} e_{k,m} x_{k,m} f_{k,m}^2, \tag{8.7}$$

where $\alpha_{k,m}$ is the unit cost for collecting each image, $x_{k,m}$ is the amount of images, and $f_{k,m}$ is the amount of computation resources. E_k is regarded as a constant for all the vehicular clients [28, 31]. According to [224], $e_{k,m} = \iota_{k,m} b_{k,m} \eta_{k,m} \rho_{k,m}$ where $\iota_{k,m}$ is the unit cost for the computation resource consumption, $b_{k,m}$ is the size of each image, $\eta_{k,m}$ is the effective switched capacitance that depends on the chip architecture, and $\rho_{k,m}$ is the number of CPU cycles to process one bit. We consider a special case that $\alpha_{k,m} = \mu_k e_{k,m}$, where μ_k could be identical for all the vehicular clients. The cost of vehicular client m is simplified into

$$c_{k,m} = \mu_k e_{k,m} x_{k,m} + E_k e_{k,m} x_{k,m} f_{k,m}^2. \tag{8.8}$$

With a lower $e_{k,m}$, vehicular client m can be more suitable to provide computation resources at a lower cost. Thus, $e_{k,m}$ is a key factor of the computation capability of vehicular client m.

Utility Function and Type of Vehicular Client

The utility of vehicular client m is related to the difference between its valuation and cost. Using (8.6) and (8.8), the utility of vehicular client m is shown by

$$u_{k,m} = \beta_{k,m} h_k(p_{k,m}) - \mu_k e_{k,m} x_{k,m} - E_k e_{k,m} x_{k,m} f_{k,m}^2. \tag{8.9}$$

To formulate the type of vehicular client m, we first transform (8.9) with $e_{k,m}$ as follows

$$\widehat{u_{k,m}} = \frac{u_{k,m}}{e_{k,m}} = \frac{\beta_{k,m}}{e_{k,m}} h_k(p_{k,m}) - \mu_k x_{k,m} - E_k x_{k,m} f_{k,m}^2. \tag{8.10}$$

The authors in [223] has claimed that the transformation has no impact on contract design. We will discuss the details later. The type of vehicular client m is represented by $\theta_{k,m} = \frac{\beta_{k,m}}{e_{k,m}}$.

Definition 1 In the k-th global iteration, the types of vehicular clients are sorted in an ascending order and classified into $\theta_{k,1}, \ldots \theta_{k,N}$, which follows

$$\theta_{k,1} < \ldots < \theta_{k,n} < \ldots < \theta_{k,N}, N \leq M. \tag{8.11}$$

The higher order of θ implies that they have greater availability to contribute their images and computation resources in the local DNN model training. Each vehicular client can easily determine its own type by measuring its image quality and computation capability while the central server is totally not aware of their exact types. But the central server can only obtain the number of each type vehicular clients through observing their historical records. Let $M_{k,n}$ represent the number of vehicular clients belonging to type-n in the k-th global iteration. We have $\sum_{n \in N} M_{k,n} = M_k$. The utility of type-$n$ vehicular client is expressed by

$$\widehat{u_{k,n}} = \theta_{k,n} h_k(p_{k,n}) - c_{k,n}(x_{k,n}, f_{k,n}), \tag{8.12}$$

where $c_{k,n}(x_{k,n}, f_{k,n}) = \mu_k x_{k,n} + E_k x_{k,n} f_{k,n}^2$.

Utility Function of Central Server

In a certain global iteration, the utility of the central server is calculated by

$$U_k = R_k - C_k, \tag{8.13}$$

where C_k is the cost function in terms of rewards, and R_k is the revenue function in terms of images and computation resources. The revenue function R_k is shown by

$$R_k = \psi_k A_k, \tag{8.14}$$

where A_k indicates the learning efficiency and ψ_k is the unit revenue for the learning efficiency. According to [220], the learning efficiency is modeled as

$$A_k = \frac{\Delta F_k}{t_k}, \tag{8.15}$$

where ΔF_k is the global loss decay, and t_k is the end-to-end latency of federated learning in one global iteration.

Global Loss Decay

According to [220], vehicular clients contribute more training images for federated learning, which results in a much lower global loss attenuation. Thus, the relationship between the global loss decay and the total amount of contributed training images can be approximately evaluated as

$$\Delta F_k = \xi \sqrt{E_k \sum_{n \in N} M_{k,n} x_{k,n}}, \tag{8.16}$$

where ξ is the coefficient determined by the specific structure of the DNN model.

End-to-end Latency

The central server starts for model aggregation only after receiving all the updated local DNN models. In the k-th global iteration, the end-to-end latency of federated learning for N types of vehicular clients is determined by

$$t_k = \max_{n \in N} t_{k,n}, t_k \leq T_k^{max}, \tag{8.17}$$

where T_k^{max} is the synchronization latency required by the central server and $t_{k,n}$ is the end-to-end latency for type-n vehicular client in the global iteration. The end-to-end latency for type-n vehicular client is calculated by

$$t_{k,n} = t_{k,n}^d + t_{k,n}^c + t_{k,n}^u, \tag{8.18}$$

where $t_{k,n}^d$ is the latency of downloading the global DNN model, $t_{k,n}^c$ is the latency of training the local DNN model, and $t_{k,n}^u$ is the latency of uploading the updated local DNN model.

- *Global Model Download Latency:* The latency of downloading the global DNN model is

$$t_{k,n}^d = \frac{\phi_{k,n}^d}{r_{k,n}^d}, \tag{8.19}$$

   where $\phi_{k,n}^d$ is the size of the global DNN model and $r_{k,n}^d$ is the downlink rate.
- *Local Model Training Latency:* Within E_k local iterations, the number of CPU cycles for type-n vehicular client to perform x_n training images, is denoted as $E_k b_{k,n} x_{k,n} \rho_{k,n}$. Thus, the latency of training the local DNN model is

$$t_{k,n}^c = \frac{E_k b_{k,n} x_{k,n} \rho_{k,n}}{f_{k,n}}. \tag{8.20}$$

- *Updated Local Model Upload Latency:* The latency of uploading the updated local DNN model is given by

$$t_{k,n}^u = \frac{\phi_{k,n}^u}{r_{k,n}^u}, \tag{8.21}$$

where $\phi_{k,n}^u$ is the size of the updated local DNN model and $r_{k,n}^u$ is the uplink rate.

For the central server, the cost C_k is formulated as

$$C_k = \sum_{n \in N} M_{k,n} p_{k,n}. \tag{8.22}$$

The entire utility function of the central server is

$$U_k = \frac{\psi_k \xi \sqrt{E_k \sum_{n \in N} M_{k,n} x_{k,n}}}{\max_{n \in N} t_{k,n}(x_{k,n}, f_{k,n})} - \sum_{n \in N} M_{k,n} p_{k,n}. \tag{8.23}$$

8.2.3   Contract Formulation

To encourage the vehicular clients to participate in the model aggregation, the contract items need to satisfy the constraints individual rationality (IR) and incentive compatibility (IC).

Definition 2 (Individual Rationality (IR)) Vehicular clients should choose the contract items ensuring a non-negative utility, i.e.,

$$\widehat{u_n}(x_n, f_n, p_n) = \theta_n h(p_n) - c_n(x_n, f_n) \geq 0, n \in \{1, 2, \ldots, N\}. \tag{8.24}$$

The IR ensures that the reward of each vehicular client compensates the cost of resource consumption in the model aggregation. If $\widehat{u_n} \leq 0$, the vehicular client will not participate in the model aggregation, i.e., choosing the contact item ($x_n = 0, f_n = 0, p_n = 0$).

Definition 3 (Incentive Compatibility (IC)) Vehicular client m must choose the contract item (x_n, f_n, p_n) matching its own type, which can be mathematically expressed as

$$\theta_n h(p_n) - c_n(x_n, f_n) \geq \theta_n h(p_j) - c_j(x_j, f_j), n, j, \in \{1, 2, \ldots, N\}. \tag{8.25}$$

The IC constraint ensures that each vehicular client automatically chooses the contract items designed for its corresponding type.

For satisfying the constraints of IC and IR, the optimization problem of maximizing the utility of the central server is formulated as

$$\max_{(\mathbf{x},\mathbf{f},\mathbf{p})} U = \frac{\psi \xi \sqrt{E \sum_{n \in N} M_n x_n}}{\max_{n \in N} t_n(x_n, f_n)} - \sum_{n \in N} M_n p_n,$$

$$\text{s.t. } C1 : \theta_n h(p_n) - c_n(x_n, f_n) \geq 0, n \in \{1, 2, \ldots, N\},$$

$$C2 : \theta_n h(p_n) - c_n(x_n, f_n) \geq \theta_n h(p_j) - c_j(x_j, f_j), n, j \in \{1, 2, \ldots, N\},$$

$$C3 : 0 \leq p_n, 0 \leq x_n, 0 < f_n, n \in \{1, 2, \ldots, N\},$$

$$\tag{8.26}$$

where C1 and C2 are IR and IC, respectively, C3 ensures decision variables are non-negative and $\mathbf{p}, \mathbf{x}, \mathbf{f} \in \mathbb{R}^N$ are vectors.

8.2.4 Problem Relaxation and Transformation

Relaxing Constraint

It is hard to solve the optimization problem in (8.26) with non-convex objective function and constraints. To make it better tractable, a new variable T is introduced to denote the end-to-end latency, i.e., $T = \max_{n \in N} t_n(x_n, f_n)$. The optimization problem in (8.26) is transformed into

$$\max_{(\mathbf{x},\mathbf{f},\mathbf{p},T)} U = \frac{\psi \xi \sqrt{E \sum_{n \in N} M_n x_n}}{T} - \sum_{n \in N} M_n p_n,$$

s.t. $C1 : \theta_n h(p_n) - c(x_n, f_n) \geq 0, n \in \{1, 2, \ldots, N\}$,

$\quad\;\; C2 : \theta_n h(p_n) - c(x_n, f_n) \geq \theta_n h(p_j) - c(x_j, f_j), n, j \in \{1, 2, \ldots, N\}$,

$\quad\;\; C3 : 0 \leq p_n, 0 \leq x_n, 0 < f_n, n \in \{1, 2, \ldots, N\}$,

$\quad\;\; C4 : \max_{n \in N} t_n = T$,

$\quad\;\; C5 : 0 < T \leq T^{max}$,

$$(8.27)$$

where C5 ensures the end-to-end latency can not exceed the synchronization latency required by the central server.

Lemma 1 *When ρ, b, t^d and t^u are constants with the same value for all vehicular clients, $\max_{n \in N} t_n = T$ is relaxed into $t^d + t^u < T$ and $f_n = \lambda(T) x_n, n \in \{1, 2, \ldots, N\}$ where $\lambda(T) = \frac{\rho b E}{(T - t^d - t^u)}$.*

Proof $\max_{n \in N} t_n = T$ is firstly relaxed into $t_n = T, n \in \{1, 2, \ldots, N\}$. $t_n = T$ is rewritten as $f_n = x_n \frac{\rho_n b_n E}{(T - t_n^d - t_n^u)}$. Referring to [225, 226], ρ_n and b_n are simplified into constants ρ and b with the same value for all vehicular clients. Similar to [28, 31], $\forall n \in N, t_n^d$ and t_n^u are set constants with the same value for all vehicular clients. As a result, $f_n = x_n \frac{\rho_n b_n E}{(T - t_n^d - t_n^u)}$ is simplified into $f_n = x_n \frac{\rho b E}{(T - t^d - t^u)}$. We define $\lambda(T) = \frac{\rho b E}{(T - t^d - t^u)}$ where $T - t^d - t^u > 0$. $f_n = x_n \frac{\rho b E}{(T - t^d - t^u)}$ is rewritten as $f_n = \lambda(T) x_n$ and $t^d + t^u < T$.

To simplify the expression, $\lambda(T)$ is expressed as λ. Replacing $\max_{n \in N} t_n = T$ in (8.27) with $f_n = \lambda x_n, n \in \{1, 2, \ldots, N\}$ and $t^d + t^u < T$, the optimization problem (8.27) is rewritten as

$$\max_{(\mathbf{x},\mathbf{f},\mathbf{p},T)} U = \frac{\psi \xi \sqrt{E \sum_{n \in N} M_n x_n}}{T} - \sum_{n \in N} M_n p_n,$$

s.t. $C1 : \theta_n h(p_n) - c_n(f_n, x_n) \geq 0, n \in \{1, 2, \ldots, N\}$,

$\quad\;\; C2 : \theta_n h(p_n) - c_n(f_n, x_n) \geq \theta_n h(p_j) - c_j(f_j, x_j), n, j \in \{1, 2, \ldots, N\}$,

$\quad\;\; C3 : 0 \leq x_n, 0 < f_n, 0 \leq p_n, n \in \{1, 2, \ldots, N\}$,

$\quad\;\; C6 : f_n = \lambda x_n, n \in \{1, 2, \ldots, N\}$,

$\quad\;\; C7 : t^d + t^u < T \leq T^{max}$,

$$(8.28)$$

where C6 and C7 comes from C4 and C5 with Lemma 1.

By replacing $f_n, n \in \{1, 2, \ldots, N\}$ in (8.28) with $f_n = \lambda x_n, n \in \{1, 2, \ldots, N\}$, we can rewrite (8.28) as

$$\max_{(\mathbf{x}, \mathbf{p}, T)} U = \frac{\psi \xi \sqrt{E \sum_{n \in N} M_n x_n}}{T} - \sum_{n \in N} M_n p_n,$$

s.t. $C1 : \theta_n h(p_n) - c_n(\lambda x_n, x_n) \geq 0, n \in \{1, 2, \ldots, N\}$,

$\quad C2 : \theta_n h(p_n) - c_n(\lambda x_n, x_n) \geq \theta_n h(p_j) - c_j(\lambda x_j, x_j), n, j \in \{1, 2, \ldots, N\}$,

$\quad C3 : 0 \leq x_n, 0 \leq p_n, n \in \{1, 2, \ldots, N\}$,

$\quad C7 : t^d + t^u < T \leq T^{max}$.

$$(8.29)$$

Using Lemma 1, (x, f, p) is simplified into $(x, \lambda x, p)$, which implies that the amount of computation resources relies the amount of images. In other words, for type-n vehicular client, type $\theta_n = \frac{\beta_n}{e_n}$ is simplified into $\theta_n = \frac{\beta_n}{e}$ which only depends the image quality.

Simplifying Complicated Constraint

Non-convex and couple constraints in (8.29), i.e., N IR constraints and $N(N-1)$ IC constraints, makes (8.29) hard to be solved directly. To reduce constraints of (8.29), we introduce the following lemmas.

Lemma 2 *Given T, for any feasible contact $(x_n, \lambda x_n, p_n)$, $p_n \geq p_j$ if and only if $x_n \geq x_j, \forall n, j \in \{1, \ldots, N\}$.*

Proof We bring the $f_n = \lambda x_n$ into the $c_n(x_n, f_n)$ given by

$$c_n(x_n, \lambda x_n) = \mu x_n + E \lambda^2 x_n^3.$$

$$(8.30)$$

It is obvious that $c_n(x_n, \lambda x_n)$ is a convex function in terms of x_n. To simplify the expression, $c_n(x_n, \lambda x_n)$ is expressed as c_n. First, we prove that if $x_n > x_j$, then $p_n > p_j$. According to constraint (8.25), we have the following inequality:

$$c_n - c_j < \theta_n(h(p_n) - h(p_j)), n, j \in N.$$

$$(8.31)$$

Since $x_n > x_j$, we can obtain $c_n - c_j > 0$. Then, $h(p_n) - h(p_j) > 0$ is satisfied. Due to the increasing valuation function of $h(\cdot)$, we have $p_i > p_j$. Furthermore, we prove that if $p_n > p_j$, then $\theta_n > \theta_j$. Referring to constraint (8.25), we have the following inequality:

$$\theta_j(h(p_n) - h(p_j)) < c_n - c_j, n, j \in N.$$

$$(8.32)$$

Since $p_i > p_j$ and $h(\cdot)$ is a monotonically increasing valuation function in terms of p, we have $\theta_j(h(p_n) - h(p_j)) > 0$. Thus, we can obtain $c_n - c_j$, i.e., $x_n > x_j$. Finally, we prove that $x_n = x_j$ if and only if $p_n = p_j$, $\forall n, j \in \{1, \ldots, N\}$. We use the similar procedure to prove $x_n = x_j$ if and only if $p_n = p_j$.

From Lemma 2, vehicular clients contribute more images resulting in more computation resources, the vehicular client will receive more reward. If two vehicular clients contribute the same amount of images, they will receive the same reward. Using Lemma 2, we can deduce Lemma 3.

Lemma 3 (Monotonicity) *Given T, for any feasible contact $(x_n, \lambda x_n, p_n)$, $p_n \geq p_j$ if and only if $\theta_n \geq \theta_j$, $\forall n, j \in \{1, \ldots, N\}$.*

Proof Following [216], we prove the sufficiency at first: if $\theta_n \geq \theta_j$, then $p_n \geq p_j$. Based on the IC constraints of type θ_n and type θ_j vehicular clients, we have

$$\theta_n h(p_n) - c_n \geq \theta_n h(p_j) - c_j, \tag{8.33}$$

and

$$\theta_j h_j - c_j \geq \theta_j h(p_n) - c(x_n). \tag{8.34}$$

Adding (8.33) and (8.34), and by rearranging, we can get $(\theta_n - \theta_j)(h(p_n) - h(p_j)) \geq 0$. As $\theta_n \geq \theta_j$, we must have $h(p_n) - h(p_j) \geq 0$. Since $p_n \geq p_j$ and $h(\cdot)$ is a monotonically increasing valuation function in terms of p, we have $p_n \geq p_j$. Next, we prove the necessity: if $p_n \geq p_j$, then $\theta_n \geq \theta_j$. Similar to the above process, we use the IC constraint to obtain the same result $(\theta_n - \theta_j)(h(p_n) - h(p_j)) \geq 0$. The reason is similar to the sufficiency.

Lemma 3 indicates that a higher type of vehicular client should get more reward, which is the monotonicity property of the contract design.

Based on the above analysis, the IC constraints are used to reduce the IR constraints. Thus, we have the following lemma.

Lemma 4 *Given T, with the IC condition, the IR constraints can be reduced as*

$$\theta_1 h(p_1) - c_1(\lambda x_1, x_1) \geq 0. \tag{8.35}$$

Proof Given that $\theta_1 < \theta_2 < \ldots < \theta_N$, we utilize IC constraints to have

$$\theta_n h(p_n) - c_n \geq \theta_n h(p_1) - c_1 \geq \theta_1 h(p_1) - c_1 \geq 0. \tag{8.36}$$

(8.36) indicates that the first type of vehicular client satisfies the IR constraint, other types of vehicular clients will satisfy the other IR constraints automatically. Thus, we need to keep the IR constraint for the first type and the other IR constraints can be reduced.

Based on the IC constaints, we also have the following lemma.

Lemma 5 *Given T, by utilizing the monotonicity in Lemma 3, the IC condition can be transformed into the Local Downward Incentive Compatibility (LDIC) given by*

$$\theta_n h(p_n) - c_n(\lambda x_n, x_n) \geq \theta_n h(p_{n-1}) - c_{n-1}(\lambda x_{n-1}, x_{n-1}), n \in \{2, \ldots, N\},$$
(8.37)

and the local upward incentive compatibility (LUIC) given by

$$\theta_n h(p_n) - c_n(\lambda x_n, x_n) \geq \theta_n h(p_{n+1}) - c_{n+1}(\lambda x_{n+1}, x_{n+1}), n \in \{1, \ldots, N-1\}.$$
(8.38)

Proof The IC constraints between types n and j, $n, j \in \{2, \ldots, N\}$ are defined as downward incentive constraints (DICs) represented as

$$\theta_n h(p_n) - c_n \geq \theta_n h(p_j) - c_j, \ \forall n, j \in \{2, \ldots, N\}, n > j.$$
(8.39)

The IC constraints between type n and type j, $n, j \in \{2, \ldots, N\}$ are defined as upward incentive constraints (UICs) represented as

$$\theta_n h(p_n) - c_n \geq \theta_n h(p_j) - c_j, \ \forall n, j \in \{2, \ldots, N\}, n < j.$$
(8.40)

Specifically, two adjacent types in UICs are defined as LUICs and two adjacent types in DICs are defined as LDICs. The LUICs and LDICs can be represented as, respectively,

$$\theta_n h(p_n) - c_n \geq \theta_n h(p_{n+1}) - c_{n+1}, \ \forall n \in \{1, \ldots, N-1\},$$
(8.41)

and

$$\theta_n h(p_n) - c_n \geq \theta_n h(p_{n-1}) - c_{n-1}, \ \forall n \in \{2, \ldots, N\}.$$
(8.42)

With the following proof, we will first reduce the DIC to the LDIC. Adopting the LDIC with three continuous types of the vehicular clients, $\theta_{n-1} \leq \theta_n \leq \theta_{n+1}, n \in \{2, \ldots, N-1\}$, we have the following inequalities

$$\theta_{n+1} h(p_{n+1}) - c_{n+1} \geq \theta_{n+1} h(p_n) - c_n,$$
(8.43)

$$\theta_n h(p_n) - c_n \geq \theta_n h(p_{n-1}) - c_{n-1}.$$
(8.44)

According to the monotonicity, i.e., $p_n > p_j$ if and only if $\theta_n > \theta_j$, we have

$$\theta_{n+1}(h(p_n) - h(p_{n-1})) \geq \theta_n(h(p_n) - h(p_{n-1})).$$
(8.45)

Combining (8.44) and (8.45), we have

$$\theta_{n+1}h(p_n) - c_n \geq \theta_{n+1}h(p_{n-1}) - c_{n-1}. \tag{8.46}$$

Combining (8.43) and (8.46), we have

$$\theta_{n+1}h(p_{n+1}) - c_{n+1} \geq \theta_{n+1}h(p_{n-1}) - c_{n-1}. \tag{8.47}$$

Using (8.47), we can prove that all the DICs can hold

$$\theta_{n+1}h(p_{n+1}) - c_{n+1} \geq \theta_{n+1}h(p_{n-1}) - c_{n-1} \geq \ldots \geq \theta_{n+1}h(p_1) - c_1. \tag{8.48}$$

Hence, we use the LDICs to hold and reduce all the DICs. Using similar process, we can also prove that all the UICs can automatically hold, when the LUICs are satisfied.

Using Lemma 2 to Lemma 5, we reduce the complicated IR and IC constraints. The optimization problem in (8.29) can be further transformed as follows

$$\max_{(\mathbf{x},\mathbf{p},T)} U = \frac{\psi\xi\sqrt{E\sum_{n\in N} M_n x_n}}{T} - \sum_{n\in N} M_n p_n,$$

$$\text{s.t. } C1 : \theta_n h(p_n) - c_n(\lambda x_n, x_n) \geq 0, n \in \{1, 2, \ldots, N\},$$

$$C3 : 0 \leq x_n, 0 \leq p_n, n \in \{1, 2, \ldots, N\},$$

$$C7 : t^d + t^u < T \leq T^{max}, \tag{8.49}$$

$$C8 : \theta_n h(p_n) - c_n(\lambda x_n, x_n) \geq$$

$$\theta_n h(p_{n-1}) - c_{n-1}(\lambda x_{k,n-1}, x_{k,n-1}), n \in \{2, \ldots, N\},$$

$$C9 : \theta_n h(p_n) - c_n(\lambda x_n, x_n) \geq$$

$$\theta_n h(p_{n+1}) - c_{n+1}(\lambda x_{n+1}, x_{n+1}), n \in \{1, 2, \ldots, N-1\},$$

$$C10 : p_1 \leq p_2 \leq \cdots \leq p_N,$$

where C8 and C9 are the LDIC and LUIC, respectively, and C10 is the monotonicity property of the contract design. Using the LDIC and the LUIC in (8.49), we deduce Lemma 6.

Lemma 6 *Given T, since the objective function of (8.49) is an increasing function in terms of x_n as well as a decreasing function of p_n, $\forall n \in \{1, \ldots, N\}$, the optimization problem in (8.49) can be further simplified as*

$$\max_{(\mathbf{x}, \mathbf{p}, T)} U = \frac{\psi \xi \sqrt{E \sum_{n \in N} M_n x_n}}{T} - \sum_{n \in N} M_n p_n,$$

s.t. $C3 : 0 \leq x_n, 0 \leq p_n, n \in \{1, 2, \ldots, N\}$,

$\quad C7 : t^d + t^u < T \leq T^{max}$,

$\quad C10 : p_1 \leq p_2 \leq \cdots \leq p_N$,

$\quad C11 : \theta_1 h(p_1) - c_1(\lambda x_1, x_1) = 0$,

$\quad C12 : \theta_n h(p_n) - c_n(\lambda x_n, x_n) = \theta_n h(p_{n-1}) - c_{n-1}(\lambda x_{n-1}, x_{n-1}), n \in \{2, \ldots, N\}$,

$$(8.50)$$

where C11 and C12 come from C9 and C10.

Proof We will first prove that the reduced IR constraint $\theta_1 h(p_1) - c_1 \geq 0$ can be reduced to $\theta_1 h(p_1) - c_1 = 0$. For the reduced IR constraint, the data requester will try its best to decrease p_1 to improve the optimization objective function U until $\theta_1 h(p_1) - c_1 = 0$.

Secondly, we will prove that the LDIC can be transformed as $\theta_n h(p_n) - c_n = \theta_n h(p_{n-1}) - c_{n-1}$, which is combined with monotonicity to ensure the LUIC hold. Notice that the LDIC $\theta_n h(p_n) - c_n \geq \theta_n h(p_{n-1}) - c_{n-1}$, $\forall n \in \{2, \ldots, N\}$ will still hold if both p_n and p_{n-1} are reduced to the same amount. To maximize the optimization objective function, the data requester will decrease p_j as possible as it can until $\theta_n h(p_n) - c_n = \theta_n h(p_{n-1}) - c_{n-1}$. Notice that this process doesn't have an effect on other types LDIC. So the LDIC can be simplified as $\theta_n h(p_n) - c_n = \theta_n h(p_{n-1}) - c_{n-1}$, $\forall n \in \{2, \ldots, N\}$.

Thirdly, we will prove that if $\theta_n h(p_n) - c_n = \theta_n h(p_{n-1}) - c_{n-1}$, $\forall n \in \{2, \ldots, N\}$ and the monotonicity hold, the LUIC holds. The constraint $\theta_n h(p_n) - c_n = \theta_n h(p_{n-1}) - c_{n-1}$, $\forall n \in \{2, \ldots, N\}$ can be transformed as

$$\theta_n h(p_n) - \theta_n h(p_{n-1}) = c_n - c_{n-1}. \tag{8.51}$$

Due to the monotonicity, i.e., if $\theta_n \geq \theta_{n-1}$, then $h(p_n) \geq h(p_{n-1})$, we further have

$$\theta_n h(p_n) - \theta_n h(p_{n-1}) \geq \theta_{n-1} h(p_n) - \theta_{n-1} h(p_{n-1}). \tag{8.52}$$

Combine (8.51) and (8.52), we have

$$\theta_n h(p_n) - \theta_n h(p_{n-1}) = c_n - c_{n-1} \geq \theta_{n-1} h(p_n) - \theta_{n-1} h(p_{n-1}). \tag{8.53}$$

Equation (8.53) equally is transformed as

$$\theta_{n-1} h(p_{n-1}) - c_{n-1} \geq \theta_{n-1} h(p_n) - c_n, \tag{8.54}$$

which is exactly the LUIC condition. So we remove the LUIC from the constraints in (8.50).

8.2.5   Solution to Optimal Contracts

To quantity the analysis, we consider a case $h(p) = p$. The similar case appears in [225]. We use the method of iterating C11 and C12 constraints to obtain p_n expressed as

$$p_n = \frac{c(x_1, \lambda x_1)}{\theta_1} + \sum_{a=1}^{n} \Delta_a, \tag{8.55}$$

where $\Delta_a = \frac{c(x_a, \lambda x_a)}{\theta_a} - \frac{c(x_{a-1}, \lambda x_{a-1})}{\theta_a}$ and $\Delta_1 = 0$. By replacing p_n in (8.50) with (8.55), we can rewrite (8.50) as

$$\max_{(\mathbf{x}, T)} U = \frac{\psi \xi \sqrt{E \sum_{n \in N} M_n x_n}}{T} - \sum_{n \in N} M_n d_n c_n,$$

$$\text{s.t. } C3 : 0 \leq x_n, n \in \{1, 2, \dots, N\}, \tag{8.56}$$

$$C7 : t^d + t^u < T \leq T^{max},$$

where $d_n = \frac{M_n}{\theta_n} + \left(\frac{1}{\theta_n} - \frac{1}{\theta_{n+1}} \right) \sum_{j=n+1}^{N} M_j$ with $n < N$, $d_n = \frac{M_n}{\theta_n}$ with $n = N$, and $c_n = \mu x_n + E \lambda^2 x_n^3$.

Given T, it can be easily verified that (8.56) is a concave optimization problem. Based on the above analysis, we design Algorithm 12 as follows:

- **Step 1:** Initializing parameters such as M, N, E, b, μ, setting $i = 1$, $T = t^u + t^d + \tau$ where τ is a step size, and $U^\star = 0$.
- **Step 2:** By solving the optimization problem in (8.56) with convex optimization, we get U^i and $\mathbf{x}^i$.
- **Step 3:** If $\frac{U^i - U^\star}{U^\star} < 10^{-5}$, the algorithm goes to step 5; If $U^\star < U^i$, $U^\star$ will be replaced by U^i. Continuously, $i = i + 1$ and $T = T + \tau$ are executed.
- **Step 4:** If $T < T^{max}$, the algorithm goes to step 2. Otherwise, the algorithm returns $U^\star$, $\mathbf{x}^\star$ and $T^\star$.
- **Step 5:** Based on $\mathbf{x}^\star$ and $T^\star$, we compute the optimal price $\mathbf{p}^\star$ and amount of images $\mathbf{x}^\star$ using (8.55) and amount of computation resources $\mathbf{f} = \lambda(T)\mathbf{x}$, respectively. Finally, the algorithm outputs $\mathbf{p}^\star$, $\mathbf{x}^\star$, $\mathbf{f}^\star$ and $T^\star$.

Algorithm 12 Contract optimization based Greedy method

1: Set $i = 1$, $T = t^d + t^u + \tau$ and $U^\star = 0$;
2: **while** $T \leq T^{max}$ **do**
3:     Get U^i with solving optimization problem (8.56) with standard convex optimization tools;
4:     **if** $\frac{U^i - U^\star}{U^\star} < 10^{-5}$ **then**
5:         **Break for**
6:     **end if**
7:     **if** $U^\star < U_i$ **then**
8:         $U^\star = U_i$,
9:     **end if**
10:     $i = i + 1$;
11:     $T = T + \tau$;
12: **end while**
13: Return $\mathbf{x}^\star$ and $T^\star$
14: Assign the optimal price $\mathbf{p}^\star$ with (8.55)
15: Compute the optimal amount of images $\mathbf{x}^\star$ with $\mathbf{f}^\star = \lambda(T^\star)\mathbf{x}^\star$
16: Return $\mathbf{p}^\star, \mathbf{x}^\star, \mathbf{f}^\star, T^\star$

Considering the implementation of Algorithm 1, we could evaluate its computational complexity, which has the form of $\mathcal{C}_x \propto N^3(T^{max} - t^d - t^u)/\tau$ and thus, $\mathcal{C}_x \sim \mathcal{O}(N^3)$. The result indicates that our approach will consume the computing resource at a moderate level for vehicular applications.

8.2.6 Numerical Results

Simulation Settings

In the simulation, the velocity of vehicular clients is set uniformly distributed in $[v^{min}, v^{max}]$, where v^{min} and v^{max} are lower and upper bounds of the velocity, respectively [227]. But the lower and upper bounds are different in urban, suburban, and highway [228]. We consider a suburban case where the velocity of vehicular clients is generated in [0,15] m/s and there are $M = 10$ vehicular clients with $N = 10$ types. By [28, 31, 210, 229], other parameters are listed in Table 8.2. We conduct the simulation in MATLAB to get the optimal contract items. The simulation experiment has two parts.

For the first part, under asymmetric information (CA), we compare the proposed selective model aggregation approach with the original FedAvg approach in terms of accuracy and efficiency of model aggregation. In the FedAvg approach, each vehicular client is supposed to have the same amount of images and randomly given computation capability. The simulation involves the public MNIST dataset [230], and the BelgiumTSC (Belgium Traffic Sign for Classification) vehicular dataset [231]. The MNIST dataset consists of 55,000 training images and 10,000 testing images of 28×28 pixels. The BelgiumTSC dataset consists of 4591 training images and 2534 testing images. Because the images in the BelgiumTSC dataset are not all

Table 8.2 Parameter setting in the simulation

Parameter	Setting
CCD pixel size	$Q = 0.011$ mm
Camera focal length	$s = 10$ mm
Perpendicular distance	$\sigma = 5$ m
Exposure time interval	$H = \frac{1}{200}$ s
Effective switched capacitance	$\eta = 10^{-28}$
Number of CPU cycles executing one bit	$\rho = 30$ cycles/bit
Unit cost for consuming computation resources	$\iota = 1$
Coefficient determined by specific structure of DNN models	$\xi = 1$
Download and upload data rate	$r^d = r^u = 6$ MB/s
Unit revenue for the learning efficiency	$\psi = 0.6, 0.8, 1.0$
Linear factor for α and e	$\mu = 5.314 \times 10^{18}$
The number of global iteration	$K = 1000$
The number of local iteration	$E = 3, 4, 5$
Parameters of the image quality	$q_1 = 0.5, q_2 = 0.8$, and $L_t = 3$

the same size, we just resize the images to a fixed size, i.e., 28×28 pixels. The comparison is divided into two cases.

- **Blurred Training Image and Unblurred Testing Image (BU)**: We randomly divide the training images into 10 groups and each group has the same amount of images. We synthesize motion-blurred images by Sun et al. [232]. The motion blur level is divided into 10 levels, i.e., $L = 1, 2, \ldots, 10$. Each group has a motion blur level. Blurred training images and unblurred testing images constitute the training and testing datasets, respectively.
- **Blurred Training Image and Blurred Testing Image (BB)**: The training dataset is produced similar to that in BU. The testing images are blurred with level $L = 3$ to constitute the testing dataset.

According to the optimal contract items designed for their own types, each vehicular client picks out a part of training images to train the local DNN model with a convolutional neural network (CNN) in PYTHON. For the MNIST dataset, the local DNN model is executed with iteration round $E = 5$ and full gradient descent. The CNN consists of two convolutional layers followed by two fully connected layers and then another 10 units activated by soft-max, with totally about 1,662,752 parameters. According to [220], the size of the local DNN model ϕ is about 6.5 MB. For the BelgiumTSC dataset, the local DNN model is executed with iteration round $E = 5$ and full gradient descent. The CNN consists of two convolutional layers followed by three fully connected layers, with totally about 274,730 parameters. The size of the local DNN model is about 1 MB.

For the second part, we firstly evaluate the optimal contract items in the CA approach. Then, we compare the utilities of the central server and the vehicular

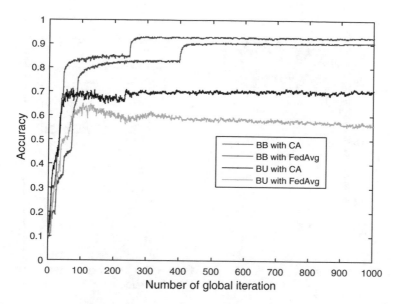

Fig. 8.5 Accuracy of model aggregation under the MNIST dataset

clients with existing baseline approaches. The first one is contract based approach under symmetric information (CS). The second one is Stackeberg game based approach under asymmetric information (SG) [214]. The third one is the linear pricing approach [215]. In the SG and the linear pricing approaches, we consider that the unit price for both images and computation resources are the same. Finally, we analyze the performance of four approaches under different system settings.

As shown in Fig. 8.5, using the MNIST dataset, we compare the accuracy of model aggregation for the CA and FedAvg approaches under BB and BU. As the number of global iteration increases, the accuracy of model aggregation is increasing for the BB and BU cases. The accuracy of model aggregation in the BB case is higher than that in the BU case. In the BB case, because the level of training image quality is closer to the level of testing image quality, which causes a high accuracy in classifying the images. In the BU case, because the gap between the level of training image quality and the level of testing image quality is large, which leads to a low accuracy in classifying the images. The similar results appear in [208]. In the BB case, the accuracy of model aggregation with the CA approach is 2.42% higher than the accuracy of model aggregation with the FedAvg approach. In the BU case, the accuracy of model aggregation adopting the CA approach is 6.28% higher than the accuracy of model aggregation adopting the FedAvg approach.

As shown in Fig. 8.6, using the BelgiumTSC dataset, we also compare the accuracy of model aggregation for the CA and FedAvg approaches under BB and BU. The accuracy of model aggregation in the BB case is also higher than that in the BU case. In the BB case, the accuracy of model aggregation with the CA approach is 1.23% higher than that of the FedAvg approach. In the BU case, the accuracy of

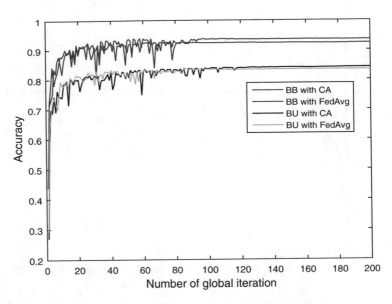

Fig. 8.6 Accuracy of model aggregation under the BelgiumTSC dataset

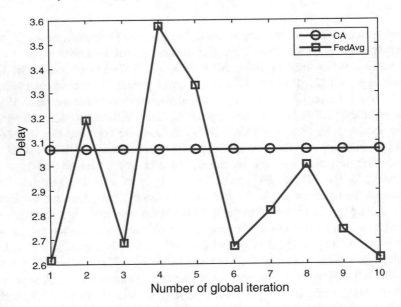

Fig. 8.7 Efficiency of model aggregation under CA and FedAvg

model aggregation adopting the CA approach is 0.2% higher than that of the FedAvg approach.

For the CA and FedAvg approaches in the MNIST dataset, Fig. 8.7 shows the efficiency of model aggregation for global iteration number $k = 1, 2, \ldots, 10$.

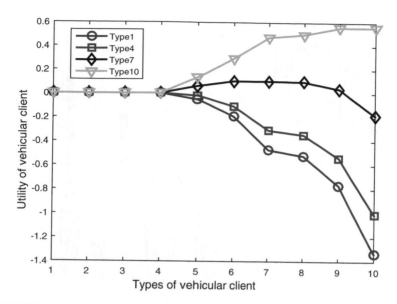

Fig. 8.8 Utilities of vehicular clients for types of vehicular clients

Since the FedAvg approach is not adapted to the random computation capability in the vehicular clients, the training latency changes in a wide range, which causes inefficient model aggregation. In the CA approach, the synchronization of training latency is beneficial for the model aggregation. The performance of efficiency of the model aggregation in the BelgiumTSC dataset has similar results to that in the MNIST.

The IR and IC constraints are verified in Fig. 8.8. It shows the utilities of type-1, type-4, type-7 and type-10 vehicular clients. The central server offers all the contract items $(p_n, x_n, f_n), n \in 1, 2, \ldots, N$ for each vehicular client. Figure 8.8 shows that the utility of each vehicular client reaches the highest when choosing the contract item designed for its own type, which satisfies the IC constraint. For instance, we consider the utility of type-7 vehicular client. If a type-7 vehicular client chooses the contract item (p_7, x_7, f_7), its utility could be maximized. Furthermore, when each vehicular client selects the contract item fitting its corresponding type, the utility of each vehicular client is nonnegative, which indicates that the IR constraint is satisfied. Therefore, after choosing the best contract item, the types of the vehicular clients will be revealed to the central server. In other words, the central server will know about the image quality and computation capability of the vehicular clients.

Figure 8.9 shows that the contract items under different types of the vehicular clients. The contract item includes the amount of images, the amount of computation resources and the reward. To show contract items in the same figure, the amount of computation resources and the reward are reduced by 10^7 times and 10^2 times, respectively. The relationship among contract items remains unchanged. As the type becomes higher, each type of vehicular client is eager to share more images and

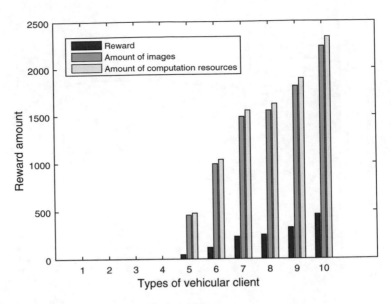

Fig. 8.9 Contract items with types of vehicular clients

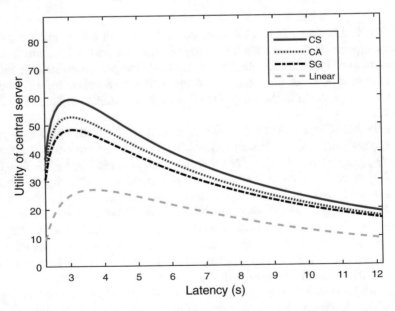

Fig. 8.10 Utility of central server under different approaches in terms of latency

computation resources for higher reward. This means Lemma 2 and Lemma 3 are both satisfied.

Figure 8.10 shows the effect of latency T on the utility of the central server under four cases, i.e., CS, CA, SG, and linear pricing approaches. We can see that as the latency grows, the utility of the central server first increases to the maximum value and then decreases. With a given latency, firstly, the CS approach achieves the best performance among four approaches and serves as upper bound. This is because the central server is fully aware of the types of the vehicular clients and tries its best to extract the revenue from the vehicular clients until the utilities of all the vehicular clients are zero. Secondly, two contract based approaches (CS and CA) have more utility at the central server than the SG approach. The contract based approaches try to extract the revenue from the vehicular clients as much as possible while satisfying both the IR and IC constraints, which will leave a small of portion of revenue for the vehicular clients. In contrast, the SG approach aims at maximizing both the utilities of the central server and the vehicular clients, which can reserve more revenue for the vehicular clients. Finally, the utility of the central server achieved by the SG approach is better than the linear pricing approach. In other words, the linear pricing approach achieves the worst performance among four approaches and serves as lower bound. This is because the linear pricing approach would not allow the central server to adapt to the change of the amount of images and computation resources, and thus make the performance become worse.

8.3 Autonomous Driving Cars

Autonomous driving cars will enable various smart features by using emerging technologies, such as edge computing, network function virtualization, cloud computing, advanced artificial intelligence schemes, 5G and beyond. The prominent features offered by autonomous driving cars are traffic sign detection, lane departure warning, collision avoidance, and instant car accident reporting, among others [233]. Additionally, autonomous driving cars can offer infotainment services based on intelligent caching [10]. One can use centralized machine learning for making autonomous driving car applications smart. However, it will suffer from the issue of data privacy leakage. Furthermore, autonomous cars generate 4000 gigaoctet of data every day [234]. Transferring the whole data to the centralized server for training a model based on centralized machine learning is difficult due to communication resource constraints. To address these challenges, we can use federated learning that enables on-device learning without the need to transfer the end-devices data to the centralized server. Instead, only learning model updates are transferred to the centralized aggregation server. It must be noted that federated learning itself has some privacy concerns and does not completely guarantee privacy preservation. Although federated learning can enable on-device machine learning for autonomous cars, it faces few challenges.

- Traditional federated learning is based on a centralized server for global aggregation and thus suffers from a robustness issue. The centralized aggregation stops working either due to a failure or security attack.

- Autonomous cars have high mobility which induces challenges for enabling frequent seamless connectivity between the end-devices and roadside units for federated learning.
- Federated learning is based on continuous interaction between the end-devices and aggregation server that will use a significant amount of communication resources. Therefore, we must efficiently perform resource management for federated learning.

To address the above challenges, one can use dispersed federated learning. In this section, we formulate an optimization problem to minimize the packet error rate, transmission latency, and transmission energy for dispersed federated learning by jointly optimizing resource allocation and association. The main contributions of this section are as follows.

- We formulate an optimization problem to minimize the packet error rate, transmission latency, and transmission energy for dispersed federated learning-enabled autonomous cars.
- Due to the NP-hard nature of the formulated problem, we decompose the main problem into two sub-problems, such as resource allocation sub-problem and association sub-problem. For the resource allocation sub-problem, we use a one-sided one-to-one matching game, whereas, for an association, we use a heuristic algorithm.
- Finally, we provide simulation results to show the effectiveness of the proposed solution.

We proposed a dispersed federated learning framework for autonomous cars in our previous work, as shown in Fig. 8.11 [235]. An optimization problem was formulated to jointly minimize the transmission latency and packet error rate by optimizing transmit power, association, and resource allocation. Block Successive Upper-Bound Minimization (BSUM)-a based solution was proposed in [235]. Although the BSUM-based solution of [235] can offer an attractive solution, it suffers from rounding errors for resource allocation and association.

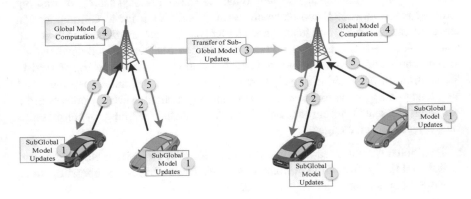

Fig. 8.11 Dispersed federated learning enabled autonomous driving cars

8.3.1 System Model and Problem Formulation

We consider a system model that consists of a set $\mathcal{M}$ of M edge computing server-based RSUs and a set $\mathcal{N}$ of N autonomous driving cars. Moreover, the set of Y cellular users served by BS in the same area as that of the RSUs and autonomous cars is denoted by $\mathcal{Y}$. In our system model, we consider set $\mathcal{R}$ of R orthogonal resource blocks that are already occupied by cellular users, for autonomous driving cars. A set $\mathcal{U}_n$ of U_n users at the autonomous car n with local datasets want to train global federated learning model.

Federated Learning Model

In federated learning, a set of devices in autonomous cars first computes their local models and send them to the aggregation server. After global aggregation, the global model is sent back to the end-devices. Every device u_n within autonomous car n has a local dataset $\mathcal{D}_u^n = [d_{u_1}^n, d_{u_2}^n, \ldots, d_{u_{k_u^n}}^n]$, where k_u^n represents the total number of data samples in the local dataset of device u_n. The size of input sample $d_{u_k}^n$ and its output $\Theta_{u_k}^n$ depend on the type of federated learning task. The output $\Theta_{u_k}^n$ is determined by weights w_u^n and the input $d_{u_k}^n$. All the users are assumed to have different dataset sizes and distribution to reflect practical scenarios [15]. The goal of the federated learning is to minimize the loss function f, i.e.,

$$\min_{w_1^n, w_2^n, \ldots, w_{U_n}^n} \frac{1}{K} \sum_{n=1}^{N} \sum_{u=1}^{U_n} \sum_{k=1}^{k_u^n} f(w_u^n, d_{u_k}^n, \Theta_{u_k}^n), \tag{8.57a}$$

$$s.t. w_u^1 = w_u^2 = \ldots = w_u^N = z, \forall u \in \mathcal{U}_n, \forall n \in \mathcal{N}, \tag{8.57b}$$

where K and z denote the size of training data for all devices of all autonomous driving cars and global federated learning model, respectively. The loss function f is dependent on the nature of the application. For instance, for prediction it accounts for prediction error. For linear regression problem, the loss function is given by $f(w_u^n, d_{u_k}^n, \Theta_{u_k}^n) = \frac{1}{2}(d_{u_k}^n w_u^n - \Theta_{u_k}^n)^2, \forall n \in \mathcal{N}$. Constraint (9.1b) ensures that all the devices must have the same federated learning model. On the other hand, the global model update is given by:

$$z = \frac{\sum_{n=1}^{N} \sum_{u=1}^{U_n} k_u^n w_u^n}{K}. \tag{8.58}$$

Federated learning involves iterative sharing of learning model updates between the end-devices and the aggregation server over a wireless channel. The wireless channel uncertainties will degrade the federated learning performance. Similar to [29], we use the packet error rate to study the degradation in federated learning performance due to wireless channel uncertainties. The packet error rate for device

u_n of autonomous car n over resource block $x_{n,m}$ and waterfall threshold ϑ is given by Xi et al. [84]:

$$q_{u,n}^r(A, X, P) = a_{n,m} x_{n,m} \left(1 - \exp \left(\frac{-\vartheta \left(\sum_{y \in \mathcal{Y}_r} h_y^r P_y^r + \sigma^2 \right)}{p_n h_{n,m}^r} \right) \right), \qquad (8.59)$$

where $a_{n,m}$ and $x_{n,m}$ denote the autonomous car-RSU association and resource allocation, respectively. p_n denotes transmit power of device n. We use binary variable $x_{n,m}$ for showing resource block allocation to autonomous cars:

$$x_{n,m} = \begin{cases} 1, & \text{If car } n \text{ is assigned resource block } r, \\ 0, & \text{otherwise.} \end{cases} \qquad (8.60)$$

The binary variable $a_{n,m}$ denotes the association of a car n with the RSU m:

$$a_{n,m} = \begin{cases} 1, & \text{If car } n \text{ is associated with RSU } m, \\ 0, & \text{otherwise.} \end{cases} \qquad (8.61)$$

The maximum number of cars associated with a RSU m must not exceed a maximum limit Δ_m, i.e.,

$$\sum_{n=1}^{N} a_{n,m} \leq \Delta_m, \forall m \in \mathcal{M}, \qquad (8.62)$$

Depending on the packet error rate, one might not consider a certain packet carrying local learning model in global aggregation due to the high packet error rate. A binary variable Π_u^n is used to indicate whether the received local model from device u_n of the autonomous car n contains errors or not ($\Pi_u^n = 1$ if it does not contain errors and 0 otherwise). To capture the effect of packet error rate on federated learning model, (8.58) can be re-written as:

$$z = \frac{\sum_{n=1}^{N} \sum_{u=1}^{U_n} k_n^u w_u^n \Pi_u^n}{\sum_{n=1}^{N} \sum_{u=1}^{U_n} k_u^n \Pi_u^n}. \qquad (8.63)$$

Now, define the cost function q_n that accounts for the effect of a packet error rate of autonomous car n on the performance of dispersed federated learning for autonomous cars [29]:

$$q_n(A, X) = \sum_{u=1}^{U_i} q_{u,n}(A, X). \qquad (8.64)$$

Communication Model

We consider orthogonal frequency division multiple access (OFDMA) in our model. Autonomous cars will use resource blocks already in use by other cellular users, and thus there will be inference between the cellular users and autonomous cars. However, there will be no interference among autonomous cars because they use different orthogonal resource blocks. A single resource block can be assigned to a maximum of one user:

$$\sum_{n=1}^{N} x_{n,m} \leq 1, \forall r \in \mathcal{R}. \tag{8.65}$$

Every autonomous car must not get more than one resource block:

$$\sum_{r=1}^{R} x_{n,m} \leq 1, \forall n \in \mathcal{N}. \tag{8.66}$$

All the autonomous driving cars must be assigned resource blocks less than or equal to the total available resource blocks.

$$\sum_{r=1}^{R} \sum_{n=1}^{N} N x_{n,m} \leq R. \tag{8.67}$$

For the up-link channel gain $h_{n,m}^r$ between the autonomous driving cars $n \in \mathcal{N}$ and RSU m for resource block r, the signal-to-interference-plus-noise ratio (SINR) is given by:

$$\Gamma_{n,m}^r = \frac{p_n h_{n,m}^r}{\sum_{y \in \mathcal{Y}_r} h_y^r P_y^r + \sigma^2}, \tag{8.68}$$

where p_n and σ^2 transmit power and noise, respectively. The term $\sum_{y \in \mathcal{Y}_i} h_y^r P_y^r$ denote the interference due to cellular users on resource block r. The up-link achievable data rate for the autonomous car n associated with RSU m for a resource block r is given by:

$$\eta_n = \Omega_{n,m}^r \log_2(1 + \Gamma_{n,m}^r), \tag{8.69}$$

where $\Omega_{n,m}^r$ represents the bandwidth allocated to the autonomous car n associated with RSU m for a resource block r. The transmission delay occurred in sending the sub-global model updates having size $v_{n,m}$ of the autonomous car n to RSU m is given by:

$$T_n^{trans}(\boldsymbol{A}, \boldsymbol{X}) = \frac{v_{n,m} x_{n,m} a_{n,m}}{\eta_n}. \tag{8.70}$$

The energy consumption in sending the sub-global model updates from autonomous car n to the RSU m is given by:

$$E_n^{trans}(A, X) = \frac{x_{n,m} a_{n,m} v_{n,m} p_n}{\eta_n}. \tag{8.71}$$

Problem Formulation

We formulate a problem to jointly minimize the dispersed federated learning cost that considers three parameters, such as (a) transmission delay, (b) effect of packet error rate, and (c) transmission energy. The cost function C_p that counts for the effect of packet loss rate on the dispersed federated learning model accuracy is given by:

$$C_p(A, X) = \sum_{n=1}^{N} q_n(A, X). \tag{8.72}$$

The total transmission delay required for one global dispersed federated learning iteration is given by:

$$C_d(A, X) = \sum_{n=1}^{N} T_n^{trans}(A, X). \tag{8.73}$$

The total energy consumed during one global dispersed federated learning iteration is given by:

$$C_e(A, X) = \sum_{n=1}^{N} E_n^{trans}(A, X). \tag{8.74}$$

The total cost for dispersed federated learning is given by:

$$C_{DFL}(A, X) = \alpha C_p(A, X) + \beta C_d(A, X) + \gamma C_e(A, X), \tag{8.75}$$

where α, β, and γ are the constants and their sum is $\alpha + \beta + \gamma = 1$. We formulate our problem joint autonomous driving cars association and resource allocation (**P1**) problem to minimize the cost associated with dispersed federated learning as follows:

$$\textbf{P1}: \min_{A,X}\ C_{DFL}(A, X) \tag{8.76}$$

subject to:

$$\sum_{n=1}^{N} x_{n,m} \leq 1, \forall r \in \mathcal{R}, \tag{8.76a}$$

$$\sum_{m=1}^{M} a_{n,m} \leq 1, \forall n \in \mathcal{N}, \tag{8.76b}$$

$$\sum_{r=1}^{R} x_{n,m} \leq 1, \forall n \in \mathcal{N}, \tag{8.76c}$$

$$\sum_{r=1}^{R} \sum_{n=1}^{N} x_{n,m} \leq R, \tag{8.76d}$$

$$\sum_{n=1}^{N} a_{n,m} \leq \Delta_m, \forall m \in \mathcal{M}, \tag{8.76e}$$

$$a_{n,m} \in \{0, 1\} \quad \forall n \in \mathcal{N}, m \in \mathcal{M}, \tag{8.76f}$$

$$x_{n,m} \in \{0, 1\} \quad \forall n \in \mathcal{N}, m \in \mathcal{M}. \tag{8.76g}$$

Problem **P1** is to minimize the total cost of one global dispersed federated learning model computation. Constraint (8.76a) restricts the assignment of a resource block to a maximum of one autonomous car. Constraint (8.76b) shows that the association of a autonomous car can be made to a maximum of one RSU. Every autonomous driving car must be not get more than one resource block according to constraint (8.76c). Constraint (8.76d) shows that the assigned resource blocks to cars must not exceed the maximum allowed limit. The maximum number of autonomous driving cars that can be associated to a particular RSU is restricted by the constraint (8.76e). Constraints (8.76f) and (8.76g) restricts that association variable $a_{n,m}$ and resource block assignment variable $x_{n,m}$ to binary values. Problem **P1** has combinatorial nature and it becomes NP-hard for large devices and resource blocks. Therefore, we decompose the problem into two sub-problems for low complexity solution.

8.3.2   Joint Association and Resource Allocation Algorithm for DFL

In this section, we present our proposed joint user association and resource allocation algorithm to minimize the cost C_{DFL} in problem **P1**. First, we decompose problem **P1** into two sub-problems: resource allocation sub-problem **P2** and autonomous driving cars association sub-problem **P3**. For a fixed association matrix

A, sub-problem **P2** is given by

$$\textbf{P2}: \min_{\mathbf{X}} \ C_{DFL}(\mathbf{X}) \tag{8.77}$$

subject to:

$$\sum_{n=1}^{N} x_{n,m} \leq 1, \forall r \in \mathcal{R}, \tag{8.77a}$$

$$\sum_{r=1}^{R} x_{n,m} \leq 1, \forall n \in \mathcal{N}, \tag{8.77b}$$

$$\sum_{r=1}^{R} \sum_{n=1}^{N} x_{n,m} \leq R, \tag{8.77c}$$

$$x_{n,m} \in \{0, 1\} \quad \forall n \in \mathcal{N}, m \in \mathcal{M}. \tag{8.77d}$$

Problem **P2** has combinatorial nature and NP-hard for large number of cars and resource blocks. On the other hand, for sub-problem **P3** we consider a fixed resource allocation matrix X and equal power to all devices, i.e.,

$$\textbf{P3}: \min_{\mathbf{A}} \ C_{DFL}(\mathbf{A}) \tag{8.78}$$

subject to:

$$\sum_{m=1}^{M} a_{n,m} \leq 1, \forall n \in \mathcal{N}, \tag{8.78a}$$

$$\sum_{n=1}^{N} a_{n,m} \leq \Delta_m, \forall m \in \mathcal{M}, \tag{8.78b}$$

$$a_{n,m} \in \{0, 1\} \quad \forall n \in \mathcal{N}, m \in \mathcal{M}. \tag{8.78c}$$

Similar to sub-problem **P2**, sub-problem **P3** has combinatorial nature and becomes NP-hard for a large number of autonomous driving cars and RSUs. To minimize the global FL cost, we propose an iterative scheme that solves sub-problem **P2** and sub-problem **P3** in an iterative manner. The iterative approach is summarized in Algorithm 13.

Matching Game-Based Resource Allocation

Sub-problem **P2** is combinatorial in nature and becomes NP-hard for a large number of autonomous driving cars and resource blocks. To solve **P2** sub-problem, we use a

---

Algorithm 13 Joint association and resource allocation algorithm for DFL

1: **Inputs**
2: Autonomous driving cars set $\mathcal{N}$, RSUs set $\mathcal{M}$, resource blocks set $\mathcal{R}$, $\Delta_m \forall m \in \mathcal{M}$, scaling constants (α, β, γ),
3: **Outputs**
4: Association matrix A, resource blocks matrix X
5: **Initialization**
6: Assignment of values to α, β, γ, and Δ_m
7: Initial random assignment of A.
8: **repeat**
9:     ***Resource Allocation***
10:    Run Algorithm 14 to yield X for fix A.
11:    ***Autonomous Car-RSUs Association***
12:    Run Algorithm 15 to yield A for fix X.
13:    Compute C_{DFL} for X and A of steps 9 and 11.
14: **until** C_{DFL} converges.

Algorithm 14 Matching-based resource allocation algorithm

1: **Inputs**
2: Resource blocks preference profile $\mathcal{R}_r$, $\forall r \in \mathcal{R}$
3: Resource blocks set $\mathcal{R}$, autonomous driving cars set $\mathcal{N}$
4: **Output**
5: Matching function $\Psi^{(t)}$
6: ***Step 1: Initialization***
7: $\mathcal{R}_r^{(t)} = \emptyset, t = 0$
8: $\Psi^{(t)} \triangleq \{\Psi(r)^{(t)}, \Psi(n)^{(t)}\}_{r \in \mathcal{R}, n \in \mathcal{N}} = \emptyset$
9: ***Step 2: Matching phase***
10: **repeat**
11:    $t \leftarrow t + 1$
12:    **for** $r \in \mathcal{R}$, propose n according to $\mathcal{R}_r^{(t)}$ **do**
13:        **if** $n \succ_r \Psi(r)^{(t)}$ **then**
14:            $\Psi(r)^{(t)} \leftarrow \Psi(r)^{(t)} \setminus n'$
15:            $\Psi(r)^{(t)} \leftarrow n$
16:            $\mathcal{R}_r^{\prime(t)} = \{n' \in \Psi(r)^{(t)} | n \succ_r n'\}$
17:        **else**
18:            $\mathcal{R}_r^{\prime\prime(t)} = \{n \in \mathcal{N} | \Psi(r)^{(t)} \succ_r n\}$
19:        **end if**
20:        $\mathcal{R}_r^{(t)} = \{\mathcal{R}_r^{\prime(t)}\} \cup \{\mathcal{R}_r^{\prime\prime(t)}\}$
21:        **for** $l \in \mathcal{R}_r^{(t)}$ **do**
22:            $\mathcal{R}_r^{(t)} \leftarrow \mathcal{R}_r^{(t)} \setminus \{l\}$
23:        **end for**
24:    **end for**
25: **until** $\Psi^{(t)} = \Psi^{(t-1)}$

---

matching game-based algorithm. Our resource allocation problem is similar to one-sided matching house allocation problem represented by a tuple $(\mathcal{E}, \mathcal{H}, \mathcal{P})$ [236–238], where $\mathcal{E}$, $\mathcal{H}$, and $\mathcal{P}$ denote the set of agents, set of houses, and preferences of agents, respectively. In our problem, the set of autonomous driving cars $\mathcal{N}$ and resource blocks $\mathcal{R}$ are equivalent to agents and houses, respectively. A one-sided matching for our resource allocation sub-problem (P2) is the assignment of resource blocks to autonomous driving cars based on a single preference list and is defined as follows:

Definition 1 A matching Ψ is a function from the set $\mathcal{N} \cup \mathcal{R}$ into the set of elements of $\mathcal{N} \cup \mathcal{R}$ such that

(1) $|\Psi(r)| \leq 1$ and $\Psi(r) \in \mathcal{N}$,
(2) $|\Psi(n)| \leq 1$ and $\Psi(n) \in \mathcal{R} \cup \phi$,
(3) $\Psi(r) = n$ if and only if n is in $\Psi(r)$,

where $\Psi(r) = n \Leftrightarrow \Psi(n) = r$ for $\forall n \in \mathcal{N}, \forall r \in \mathcal{R}$ and $|\Psi(.)|$ denotes the cardinality of matching outcome. The intuition of properties (1) and (2) is because of the constraints (8.76c) and (8.76a) that restricts the assignment of a resource block to a maximum of one autonomous car and assignment of an autonomous car to a maximum of one resource block, respectively. There is a need to define a preference list for matching game. To do so, all the resource blocks rank the autonomous driving cars according to their cost $C_{DFL}(A)$ for a fixed association matrix A to yield a preference profile matrix $\mathcal{R}_r$. A car with lowest cost $C_{DFL}(A)$ has a highest preference, and vice versa. We use a one-sided one-to-one matching game for resource allocation due to constraints in our system model that every autonomous car can get a maximum of one resource block and every resource block must not be assigned more than one autonomous driving cars. The summary of the algorithm is given in Algorithm 14.

Definition 2 For a stable matching Ψ, it is necessary that there must not be any blocking pair (n, r), where $n \in \mathcal{N}, r \in \mathcal{R}$, such that $r \succ_n \Psi(r)$, where $\Psi(r)$ denote the existing matching pair of r.

Theorem 1 *The one-sided matching Ψ produces local sub-optimal result for resource allocation problem P2.*

Proof We prove this theorem by contradictions for given autonomous driving cars-RSU association and transmit power allocation. The outcome of Algorithm 14 is the matrix $\Psi^{(t)} \mapsto X^{(t)}$ and want to minimize the DFL cost C_{DFL}. As the matching algorithm, Algorithm 14 is based on defer/acceptance. Therefore, the matching $\Psi^{(t)}$ at iteration t guarantees $C_{DFL}^{(t)} \leq C_{DFL}^{(t-1)}$. For binary resource allocation matrix X, the cost function is characterized by a non-increasing nature. Additionally, the resource allocation might not be locally optimal. For instance, one of the cars can be at the same priority in a preference profile for different resource blocks. In such a scenario, the proposed matching algorithm will assign the car to one of the resource blocks that might not be the one for which the cost C_{DFL} is lower than the other. Therefore, matching Ψ based resource allocation converges to local sub-optimal results.

Algorithm 15 Autonomous car-RSU association algorithm

1: **Inputs**
2: Resource blocks matrix X, autonomous driving cars matrix $\mathcal{N}$, RSU matrix $\mathcal{M}$, $\Delta_m \forall m \in \mathcal{M}$,
   $t = 0$
3: **Output**
4: Association matrix A
5: **Step 1: Initialization phase**
6: Compute the matrix $C_{DFL}(X)$ $\forall n \in \mathcal{N}, m \in \mathcal{M}$
7: $G^{(0)} \leftarrow C_{DFL}$ for input X and all possible associations.
8: **Step 2: Association phase**
9: **repeat**
10:      $t \leftarrow t + 1$
11:      Compute $l^{(t)} = min(G^{(t)})$
12:      For $l^{(t)}$, propose corresponding RSU $m^{(t)}$
13:      **if** $|m^{(t)}| \leq \Delta_m$ **then**
14:          $A^{(t)}(m) \leftarrow$ corresponding n
15:          $G^{(t)}(n, :) \leftarrow \emptyset$
16:      **else**
17:          $G^{(t)}(n, m) \leftarrow \emptyset$
18:      **end if**
19: **until** All autonomous cars are associated with RSUs.

Autonomous Car-RSU Association Algorithm

Sub-problem **P3** is combinatorial in nature and is NP-hard for a large number of autonomous driving cars and RSUs. Therefore, sub-problem **P3** cannot be solved using convex optimization schemes. We propose an efficient heuristic algorithm for the association of autonomous driving cars with RSUs. The association of autonomous driving cars with the RSU for a particular resource block must be done in a way to decrease the cost of the system. Our algorithm minimizes the C_{DFL} as a cost function for a given resource assignment matrix X.

Theorem 2 *The autonomous car-RSU association using Algorithm 15 is local optimal.*

Proof Similar to Theorem 1, we prove this theorem by contradictions. Algorithm 15 uses $C_{DFL} \mapsto G^{(0)}$ as a metric for a fixed resource block matrix X to compute A. In every iteration t, a minimum value of matrix $G^{(t)}$ is computed and its corresponding autonomous car is associated with the suggested RSU while fulfilling the constraint that associated cars to an RSU must not exceed its maximum limit. Such a process of autonomous car-RSU association takes place till the association of all the cars. Moreover, the condition $min(G^{(t-1)}) \leq min(G^{(t)})$ is full filled by Algorithm 15 for all the iterations. Therefore, we can say that the autonomous car-RSU association performed by Algorithm 15 achieves locally optimal results.

The summary of the proposed association scheme is given Algorithm 15. Initially, the resource blocks assignment matrix X, autonomous driving cars matrix $\mathcal{N}$, RSUs matrix $\mathcal{M}$, and maximum cars per mth miner δ_m, is fed as an input. In the *initialization phase*, the matrix $G^{(0)}$ is computed using C_{DFL} for input X for all possible associations between autonomous driving cars and RSUs (lines 6–7). In the *association phase* (lines 10–18), the first $l^{(t)}$ is computed by finding minimum cost value among all the elements of the matrix $G^{(t)}$. This value of $l^{(t)}$ represents the optimal association among all the possible associations. After fulfilling the condition of the maximum number of autonomous driving cars per RSU, the rows of the matrix $G^{(t)}$ corresponding to the later associated autonomous driving cars are deleted to enable easier computation of the minimum value (line 11) in the next iteration. Such an iterative process continues until all the autonomous driving cars are associated with the RSUs.

8.3.3 Numerical Results

In this section, we present numerical results to validate our proposal for various simulation scenarios. The LTE-based vehicular network is used for analysis, as shown in Fig. 8.12 [240–242]. We consider an area of $500 \times 500\,\mathrm{m}^2$ where a BS and 6 RSUs are deployed at the center and distributed uniformly, respectively. The cellular users and autonomous driving cars (on roads only) 30 each, are deployed randomly. Other simulation details are given in Table 8.3. Furthermore, all the values are computed using an average of 500 runs each with different positions of autonomous cars and cellular users. However, the position of the RSUs remains constant. C_{DFL} in our numerical results denote the average cost of the proposed DFL scheme. The word iteration used in this section is different from the local learning model iterations and it refers to Algorithm 13 iteration. Furthermore, we compare the performance of our proposed algorithm with two baseline schemes such as baseline-1 and baseline-2. Baseline-1 uses the proposed association scheme with random resource allocation, whereas baseline-2 uses a proposed resource allocation with the random association.

Consider Fig. 8.13, which shows plot of C_{DFL} vs. different values of constants (α, β, and γ) for 30 autonomous driving cars, 30 iterations, and 6 RSUs. The constants α, β, and γ in C_{DFL} scale effect due to packet error ratio on FL model

Table 8.3 Simulation parameters [239, 240]

Simulation parameter	Value
Vehicular network area	$500 \times 500\,\mathrm{m}^2$
Autonomous cars	30
Cellular users	30
Frame structure	Type 1 (FDD)
Carrier frequency (f)	2 GHz
Cars transmit power	23 dBm
Sub carriers per resource block	12
Resource block bandwidth (W)	180 kHz
Thermal noise for 1 Hz at 20. C	−174 dBm

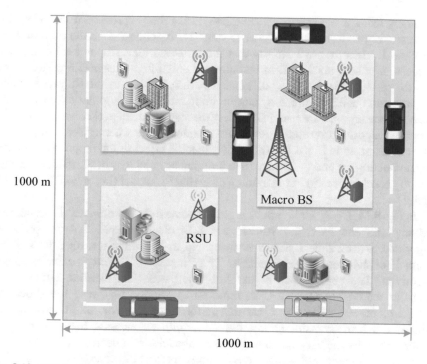

Fig. 8.12 LTE-based vehicular network layout

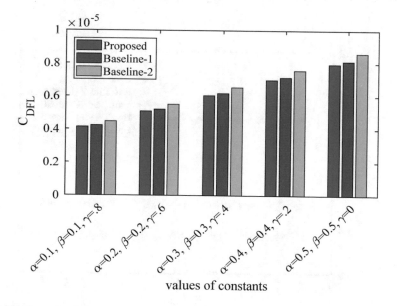

Fig. 8.13 C_{DFL} for different values of α, β and γ

accuracy, latency, and energy consumption during the FL process, respectively. For all cases, C_{DFL} has a slightly greater value for baseline-2 than the proposed and baseline-1. Therefore, we can say that the proposed joint resource allocation and association algorithm for FL in autonomous cars offers reasonable performance. For the case of $\alpha = 0.1$, $\beta = 0.1$, and $\gamma = 0.8$, the cost of the FL process takes into account the effect of energy consumption more than the other two parameters such as latency and loss in FL model accuracy due to packet error rate. The proposed algorithm results in lower cost compared to other values compared to other cases given in Fig. 8.13. This shows that the effect of the proposed algorithm on minimizing the energy is more than joint latency and loss in FL model accuracy due to packet error ratio in the training of global FL model (i.e., $\alpha = 0.5$, $\beta = 0.5$, and $\gamma = 0$).

In Fig. 8.14, C_{DFL} for iterations is plotted using different number of autonomous driving cars and fixed 6 RSUs. Figure 8.14 clearly shows that the proposed algorithm and baseline schemes converge up to 3–5 iterations regardless of the number of autonomous driving cars. This shows the stability of the proposed algorithm. The performance of baseline-1 is better than baseline-2 due to the fact that the cost of DFL depends more on association than resource allocation. Furthermore, Fig. 8.14 shows that the performance in terms of FL cost for the proposed scheme remains almost the same with an increase in the number of autonomous driving cars. On the other hand, the DFL cost decreases slightly for an increase in the number of autonomous driving cars for the brute force algorithm. The reason for this behavior is due to the fact that increasing the number of autonomous driving cars results in a high probability of getting connected to nearby RSU, and

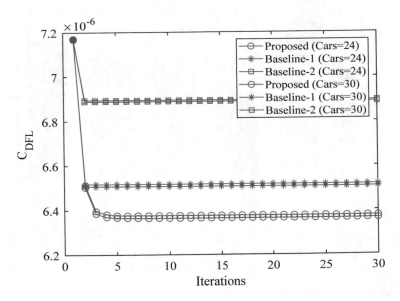

Fig. 8.14 C_{DFL} vs. iterations for $\alpha = \beta = \gamma = 1/3$ and different autonomous driving cars

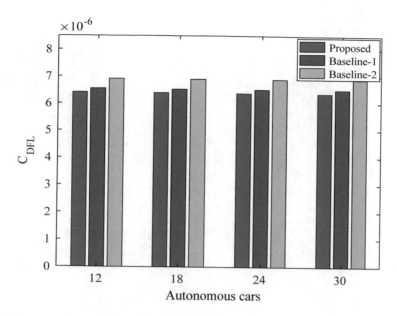

Fig. 8.15 C_{DFL} vs. autonomous driving cars for $\alpha = \beta = \gamma = 1/3$

thus offers cost reduction. The effect of an increase in the number of autonomous driving cars for a fixed number of RSUs on the C_{DFL} is shown in Fig. 8.15. Cost C_{DFL} remains almost the same with an increase in the number of autonomous driving cars for both proposed and baseline schemes. On the other hand, the effect of an increase in the number of RSUs for a fixed number of autonomous driving cars and cellular users is shown in Fig. 8.16. Cost C_{DFL} has a slightly decreasing trend for both proposed and baseline schemes with an increase in the number of RSUs for fixed autonomous cars. The reason is throughput improvement of the whole network which subsequently reduces cost C_{DFL}. Every RSU has a certain capacity to serve a maximum number of autonomous cars. Therefore, increasing the number of RSUs increases the possibility of an autonomous car getting associated with nearby RSUs than the remote one. Associating with a nearby RSU compared to remote RSU increases throughput which in turn decreases cost.

8.4 Summary

In this chapter, we have discussed the role of federated learning in enabling vehicular network applications. In the first part, we presented contract theory-enabled federated learning for vehicular networks. In the second part, we proposed a novel dispersed federated learning framework for autonomous driving cars that are based on decentralization. An optimization problem is formulated to jointly

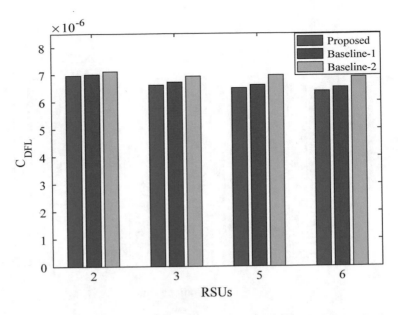

Fig. 8.16 C_{DFL} vs. autonomous driving cars for $\alpha = \beta = \gamma = 1/3$

minimize transmission latency, transmission energy, and effect of packet error rate by optimizing resource allocation and cars-RSUs association. The proposed dispersed federated learning framework for autonomous driving cars can offer robust and resource-efficient operation, and thus a promising candidate for deployment in future autonomous cars.

Chapter 9
Smart Industries and Intelligent Reflecting Surfaces

Abstract In this chapter, we present several Internet of Things applications that can leverage federated learning. More specifically, we introduce two applications such as smart industry and intelligent reflecting surfaces that can be effectively enabled by federated learning with many advantages compared to centralized machine learning. For both applications, first, we propose a framework. Then, we formulate optimization problems with their possible solutions. Finally, we provide extensive simulation results to validate our proposals.

9.1 Smart Industry

Smart industry uses collaborative robotics, edge computing, cloud computing, cyber-physical systems, cognitive Internet of things (C-IoT), and advanced machine learning schemes to enable various smart applications [243]. The devices of smart industries generate a significant amount of data that offers us the opportunity to use that data for training a machine learning model. One way can be to use centralized machine learning that is based on transferring end-devices data to the centralized server for training. However, this approach will suffer from the users' privacy leakage issue [15]. A malicious user can attack the centralized server and access the end-devices data. To address this issue in smart industries, one can use federated learning that enables on-device machine learning without the need to transfer the end-devices data to the centralized server for training. Although federated learning in smart industries can offer many benefits, it has few challenges.

- Training a federated learning model for smart industries requires a significant amount of communication resources.
- Traditional federated learning based on a single centralized server has robustness concerns in case of the centralized server failure.
- A collaborative federated learning model must take into account the validity of local learning models prior to performing aggregation. Injecting false local learning models will prolong the federated learning time.

To overcome the aforementioned challenges, one can use dispersed federated learning that can offer robust operation and efficient communication resources reuse [244]. On the other hand, the co-location of various industries is also gaining significant interest from the community. Co-location of several telecommunication industries for sharing of backup power supply for the cost-efficient operation was discussed in [245, 246]. Furthermore, various industries in a typical industrial zone are located in close vicinity. Therefore, one can use federated learning to train collaborative machine learning for smart industries. The contributions of this section are as follows.

- We present a dispersed federated learning model for smart industries that can offer a robust operation.
- An optimization problem is formulated that jointly minimizes the transmission latency and packet error rate of dispersed federated learning model computation by optimizing the resource allocation and transmit power allocation.
- Due to the NP-hard and non-convex nature of the formulated problem, we use of block successive upper-bound minimization algorithm.
- Finally, we provide numerical results to validate the proposal.

9.1.1 System Model and Problem Formulation

Consider the system model shown in Fig. 9.1 that consists of a set $\mathcal{I}$ of I industries. Within every industry there is a set $\mathcal{U}_i, \forall i \in \mathcal{I}$ of $U_i, \forall i \in \mathcal{I}$ devices with local datasets. Furthermore, there is a set $\mathcal{B}_i, \forall i \in \mathcal{I}$ of $B_i, \forall i \in \mathcal{I}$ of edge computing-based small cell base stations (SBSs) within every industry. For communication, the devices use a set $\mathcal{R}$ of R orthogonal resource blocks. To enable robust dispersed federated learning, the steps are given below.

- All the local devices compute their local learning models.
- The computed local learning models are sent to the SBS for sub-global aggregation. Prior to sub-global aggregation, the local learning models are verified by the SBSs. After sub-global aggregation, the sub-global model is sent back to the end-devices. This process will continue iteratively for a fixed number of sub-global iterations.
- After a predefined number of sub-global iterations, the sub-global model is shared among the SBSs of various industries. There can be different ways to share the sub-global models, such as direct transfer and encryption/decryption-based schemes. The encryption/decryption-based scheme can offer a more secure exchange of sub-global model updates but at the cost of computational complexity and communication overhead.
- After the transfer of sub-global model updates, global aggregation takes place. Finally, the SBS sends back the global model updates to the end-devices.

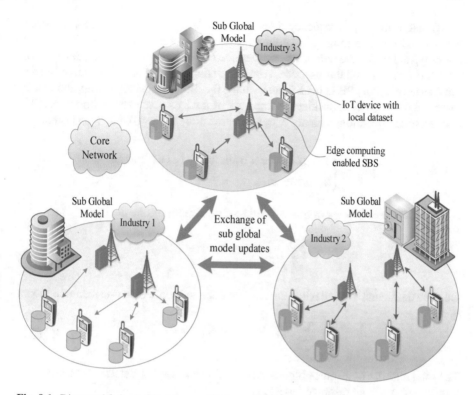

Fig. 9.1 Dispersed federated learning-enabled smart industries

Consider the devices set $\mathcal{D}_u^i = [d_{u1}^i, d_{u2}^i, \ldots, d_{uk_u^i}^i]$ of industry u, where k_u^i denotes the total number of data samples for a device u of industry i. The size of input and output for federated learning depends on the application. We consider a single output Θ_{uk}^i that is computed by w_u^i for a given input d_{uk}^i. The aim of federated learning is to minimize the loss function f.

$$\underset{w_1^i, w_2^i, \ldots, w_{U_u}^i}{\text{minimize}} \frac{1}{K} \sum_{i=1}^{I} \sum_{u=1}^{U_i} \sum_{k=1}^{k_u^i} f(w_u^i, d_{uk}^i, \Theta_{uk}^i), \tag{9.1a}$$

$$s.t. \, w_u^1 = w_u^2 = \ldots = w_u^i = z, \forall u \in \mathcal{U}_i, \forall i \in \mathcal{I}, \tag{9.1b}$$

where z and K denote the global federated learning model and the total number of data points of all devices, respectively. Constraint (9.1b) restricts the same learning model for all devices. The global model update is given by:

$$z = \frac{\sum_{i=1}^{I} \sum_{u=1}^{U_i} k_u^i w_u^i}{K}. \tag{9.2}$$

We use orthogonal frequency division multiple access (OFDMA) as an access scheme. Different orthogonal resource blocks are assigned to devices involved in learning, and thus there will be no interference between them. However, devices will interfere with the cellular users because of reusing cellular users' resource blocks. Furthermore, all the SBSs in an industry have the limited capacity to simultaneously serve the devices. We consider the fixed set of devices associated with every SBS. For resource allocation, a binary resource block allocation variable $y_{i,r}^u$ is used.

$$y_{i,r}^u = \begin{cases} 1, & \text{If device } u \text{ of industry } i \text{ is assigned } r, \\ 0, & \text{otherwise.} \end{cases} \quad (9.3)$$

A single resource block can be assigned to a maximum of one device:

$$\sum_{i \in \mathcal{I}} \sum_{u \in \mathcal{U}_i} y_{i,r}^u \le 1, \forall r \in \mathcal{R}. \quad (9.4)$$

On the other hand, every device must not be assigned more than one resource block:

$$\sum_{r \in \mathcal{R}} y_{i,r}^u \le 1, \forall i \in \mathcal{I}, u \in \mathcal{U}_i. \quad (9.5)$$

The total number of resource blocks assigned to all devices must not exceed the total number of available resource blocks:

$$\sum_{r \in \mathcal{R}} \sum_{i \in \mathcal{I}} \sum_{u \in \mathcal{U}_i} y_{i,r}^u \le R. \quad (9.6)$$

The signal-to-interference-plus-noise ratio (SINR) for a device u of industry i is given by:

$$\Gamma_r^{u \to b_i} = \frac{p_{i,u} h_{i,r}^{u \to b_i}}{\sum_{c \in \mathcal{C}_r} h_c^r P_c^r + \sigma^2}, \quad (9.7)$$

where $h_{i,r}^{u \to b_i}$ and $p_{i,u}$ denote channel gain between device u of industry i and base station b_i and the transmission power of device u of industry i, respectively. $\mathcal{C}_r$ and σ^2 represent the set of cellular users using the resource block r and noise, respectively. The term $\sum_{y \in \mathcal{C}_r} h_c^r P_c^r$ denotes the interference due to cellular users. The transmit power of all devices must be within the limits.

$$0 \le p_{i,u} \le P_m. \quad (9.8)$$

The sum of transmit power of all devices must be less than or equal the total power.

$$\sum_{i=1}^{I} \sum_{u_i=1}^{U_i} \sum_{r=1}^{R} p_{i,u} \leq P_{\max} \tag{9.9}$$

The data rate of the device u using resource block r with bandwidth $A_{u,i}^r$ is given by:

$$R_r^{u \to b_i} = A_{u,i}^r \log_2(1 + \Gamma_r^{u \to b_i}). \tag{9.10}$$

Wireless channel uncertainties significantly affect the performance of federated learning in addition to latency. The packet error rate due to a wireless channel is given by:

$$e_{u,i}(Y, P) = y_{i,r}^u \Xi, \tag{9.11}$$

where

$$\Xi = \left(1 - \exp\left(\frac{-\vartheta\left(\sum_{c \in \mathcal{C}_r} h_c^r P_c^r + \sigma^2\right)}{p_{i,u} h_i^{u \to b_i}}\right)\right), \tag{9.12}$$

Let the local learning model of device u_i of industry i consist of g_u^i bits. The total transmission time for computing the sub-global models using I_{sg} sub-global iterations is given by:

$$T_{sg}^{b_i}(Y, P) = I_{sg} \sum_{i \in \mathcal{I}} \sum_{b \in \mathcal{B}_i} \left(\sum_{u \in \mathcal{U}_i} \frac{y_{i,r}^u g_u^i}{R_r^{u \to b_i}}\right), \forall b_i \in \mathcal{B}_i. \tag{9.13}$$

In typical federated learning, an increase in local model accuracy will generally cause an improvement in global learning accuracy and vice versa. The notion of relative local accuracy is used in our model to reflect the performance of the local device on the global federated learning model. Greater the value of relative local accuracy less will be the local accuracy and vice versa. For a sub-global accuracy, ϵ and relative local accuracy θ, the number of sub-global iterations for a constant χ can be given by Konečnỳ et al. [72].

$$I_{sg}(\epsilon, \theta) = \frac{\chi \log(1/\epsilon)}{1 - \theta}. \tag{9.14}$$

Using (9.14), we can re-write (9.13) as follows.

$$\overline{T_{sg}}^{b_i}(Y, P) = \frac{T_{sg}^{b_i}(Y, P)}{1 - \theta} \tag{9.15}$$

Using Taylor's approximation, re-write (9.15) as follows.

$$\overline{T_{sg}}^{b_i}(Y, P) = (1 + \theta)(T_{sg}^{b_i}(Y, P)) \tag{9.16}$$

The effect of packet error rate on dispersed federated learning can be given by Chen et al. [29].

$$E_{sg}(Y, P) = I_{sg} \sum_{i=1}^{I} \sum_{u=1}^{U_i} e_{u,i}(X, Y). \tag{9.17}$$

Similar to (9.16), we re-write (9.17) as follows.

$$\overline{E_{sg}}(Y, P) = (1 + \theta)E_{sg}(Y, P). \tag{9.18}$$

Now we define the cost function for dispersed federated learning that jointly accounts for packet error rate and transmission latency.

$$C_{DFL}(Y, P) = \overline{T_{sg}}^{b_i}(Y, P) + \overline{E_{sg}}(Y, P). \tag{9.19}$$

Now, we formulate problem **P-1** that minimizes the cost C_{DFL} as follows:

$$\mathbf{P1} : \underset{Y,P}{\text{minimize }} C_{DFL}(Y, P) \tag{9.20}$$

subject to:

$$\sum_{i \in \mathcal{I}} \sum_{u \in \mathcal{U}_i} y_{i,r}^u \leq 1, \forall r \in \mathcal{R}, \tag{9.20a}$$

$$\sum_{r \in \mathcal{R}} y_{i,r}^u \leq 1, \forall i \in \mathcal{I}, u \in \mathcal{U}_i, \tag{9.20b}$$

$$\sum_{r \in \mathcal{R}} \sum_{i \in \mathcal{I}} \sum_{u \in \mathcal{U}_i} y_{i,r}^u \leq R, \tag{9.20c}$$

$$0 \leq p_{i,u} \leq P_m, \tag{9.20d}$$

$$\sum_{i=1}^{I} \sum_{u_i=1}^{U_i} \sum_{r=1}^{R} p_{i,u} \leq P_{\max}, \tag{9.20e}$$

$$y_{i,r}^u \in \{0, 1\} \quad \forall i \in \mathcal{I}, u \in \mathcal{U}_i. \tag{9.20f}$$

Problem **P1** is a mixed-integer non-linear programming problem. Constraints (9.20a) and (9.20b) restricts the assignment of the orthogonal resource block to a maximum of one device and a maximum of one resource block per device, respectively. Constraint (9.20c) ensures that the total number of resource blocks assigned to devices must not exceed the maximum limit of the available resource blocks. (9.20d) sets the upper and lower limit of transmit power. Constraint (9.20e) shows that the total power of all devices must not exceed the maximum available power. Finally, constraints (9.20f) restricts $y_{i,r}^u$ to be assigned only binary values.

9.1.2 Block Successive Upper-Bound Minimization-Based Solution

Due to the NP-hard nature of the formulated problem, we propose a BSUM-based scheme for a solution. To employ BSUM for cost minimization of DFL, we rewrite the optimization problem **P1** as follows.

$$\min_{\mathbf{Y}\in\mathcal{Y},\mathbf{P}\in\mathcal{P}} \mathcal{C}(\mathbf{Y},\mathbf{P}) \tag{9.21}$$

where $\mathcal{C}(\mathbf{Y},\mathbf{P}) = \mathcal{C}_{\text{DFL}}(\mathbf{Y},\mathbf{P})$. Furthermore, the feasible of sets of $\mathbf{Y}$, and $\mathbf{P}$ are:

$$\mathcal{Y} \triangleq \{\mathbf{Y} : \sum_{i\in\mathcal{I}}\sum_{u\in\mathcal{U}_i} y_{i,r}^u \leq 1, \forall r \in \mathcal{R}, \sum_{r\in\mathcal{R}} y_{i,r}^u \leq 1, \forall i \in \mathcal{I}, u \in \mathcal{U}_i,$$

$$\sum_{r\in\mathcal{R}}\sum_{i\in\mathcal{I}}\sum_{u\in\mathcal{U}_i} y_{i,r}^u \leq R, y_{i,r}^u \in \{0,1\}\},$$

$$\mathcal{P} \triangleq \{\mathbf{P} : 0 \leq p_{i,u} \leq P_m, \sum_{i=1}^{I}\sum_{u_i=1}^{U_i}\sum_{r=1}^{R} p_{i,u} \leq P_{\max}\}.$$

For the indices set $\mathcal{I}$, k, $\forall i \in \mathcal{I}^k$ for every iteration. The problem in (9.21) is still non-convex even after transforming the binary resource allocation variable into a continuous variable. Therefore, the block coordinate descent(BCD) scheme cannot be applied for solving it. To address this issue, one can add a proximal upper-bound function $\mathcal{C}_i$ of the objective function. Here, we add a quadratic penalty term for penalty parameter $\mu > 0$, whose basic purpose is to maintain h convex.

$$\mathcal{C}_i(\mathbf{Y}_i; \mathbf{Y}^k, \mathbf{P}^k) = \mathcal{C}(\mathbf{Y}_i; \tilde{\mathbf{Y}}, \tilde{\mathbf{P}}) + \frac{\mu_i}{2} \| (\mathbf{Y}_i - \tilde{\mathbf{Y}}) \|^2 . \tag{9.22}$$

Algorithm 16 BSUM algorithm

1: **Initialization:** Set $k = 0$, $\epsilon_1 > 0$, and find initial feasible solutions $(\mathbf{Y}^{(0)}, \mathbf{P}^{(0)})$;
2: **repeat**
3:      Choose index set $\mathcal{I}^k$;
4:      Let $\mathbf{Y}_i^{(k+1)} \in \min_{\mathbf{Y}_i \in \mathcal{Y}} \mathcal{C}_i\left(\mathbf{Y}_i; \mathbf{Y}^{(k)}, \mathbf{P}^{(k)}\right)$;
5:      Set $\mathbf{Y}_j^{(k+1)} = \mathbf{Y}_j^k, \forall j \notin \mathcal{I}^k$;
6:      Find $\mathbf{P}_i^{(k+1)}$, by solving (9.24);
7:      $k = k + 1$;
8: **until** $\left\| \dfrac{\mathcal{C}_i^{(k)} - \mathcal{C}_i^{(k+1)}}{\mathcal{C}_i^{(k)}} \right\| \le \epsilon_1$

9: Then, set $\mathbf{Y}_i^{(k+1)}, \mathbf{P}_i^{(k+1)})$ as the desired solution.

Similarly, we can use the quadratic penalty for $\mathbf{P}_i$. Furthermore, h with respect to $\mathbf{Y}_i$, and $\mathbf{P}_i$ in every iteration in (9.22) produces unique $\tilde{\mathbf{Y}}$, and $\tilde{\mathbf{P}}$. These values can be used as solution of $(k - 1)$ iteration. For $(k + 1)$ iteration, the solution can be:

$$\mathbf{Y}_i^{(k+1)} \in \min_{\mathbf{Y}_i \in \mathcal{Y}} \mathcal{C}_i\left(\mathbf{Y}_i; \mathbf{Y}^{(k)}, \mathbf{P}^{(k)} \right), \tag{9.23}$$

$$\mathbf{P}_i^{(k+1)} \in \min_{\mathbf{P}_i \in \mathcal{P}} \mathcal{C}_i\left(\mathbf{P}_i; \mathbf{P}^k, \mathbf{X}^{(k+1)} \right), \tag{9.24}$$

To solve sub-problems in the above equations, we use Algorithm 16.

9.1.3   Simulations

This section presents numerical results to show the validity of the proposed scheme. We use two baselines for comparisons, such as baseline-R and baseline-P. Baseline-R denotes the use of proposed resource allocation with random power allocation, whereas baseline-P represents the use of proposed power allocation with random resource allocation. We consider an LTE-based network for the industry in an area of $1000 \times 1000 \, \text{m}^2$. Moreover, cellular users also exist within the same area as that of industry. Other simulation parameters are given in Table 9.1.

Consider Fig. 9.2 which shows C_{DFL} vs. iterations for various schemes, such as proposed, baseline-R, and baseline-P. The proposed scheme outperformed all the baselines. Moreover, it converges within reasonable global iterations. Therefore, we can say that the proposed BSUM-based scheme converges fast. On the other hand, the performance of baseline-P is better than the baseline-R. The reason for this is more prominent effect of power allocation on the C_{DFL} compared to resource allocation in our scenario. Figure 9.3 shows the C_{DFL} vs. devices for various schemes. The proposed scheme outperformed baselines for a different numbers

Table 9.1 Simulation parameters [244, 247]

Simulation parameter	Value
Industrial network area	$1000 \times 1000\,\text{m}^2$
Industrial devices	30
Cellular users	30
Frame Structure	Type 1 (FDD)
Carrier frequency (f)	2 GHz
Devices transmit power	23 dBm
Sub carriers per resource block	12
Resource block bandwidth (W)	180 kHz
Thermal noise for 1 Hz at 20 °C	−174 dBm

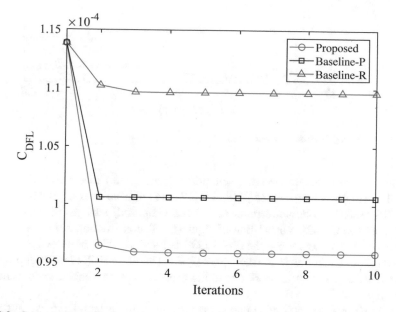

Fig. 9.2 C_{DFL} vs. iterations for various schemes

of devices. Both Figs. 9.2 and 9.3 revealed the effectiveness of our proposed cost minimization algorithm.

9.2 Intelligent Reflecting Surfaces

9.2.1 Introduction

Intelligent reflecting surface (IRS) has been proposed to improve the performance of wireless networks [248–250]. Specifically, IRS is a flat array consists of a large number of passive reflective elements, which can change the characteristics of

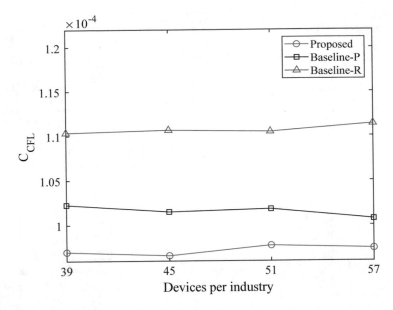

Fig. 9.3 C_{DFL} vs. devices for fixed SBSs

incident signal (e.g., frequency, amplitude or phase, etc.) to enhance the desired signal or suppress interference [248, 251]. There are a number of scenarios about the IRS assisted communications have been proposed, such as secure wireless communication [252], virtual line-of-sight (LOS) construction between the base station (BS) and users when the direct LOS is blocked [253] and device-to-device (D2D) communications in IoT networks [252]. Therefore, IRS is more appropriate for the further wireless network due to the properties of passivity and convenience of deployment.

To maximize the achievable rate at the receiver is important. Here, we introduce an approach based on deep learning (DL) to design the IRS configuration matrix (i.e., the phase shift coefficient matrix), which utilized the sampled channel state information (CSI) for IRS training. However, the protection of user's privacy during the communication process has not been fully studied in previous works as a crucial issue. In recent years, federated learning (FL) has been proposed as a new method to address the data privacy issues for distributed learning. For example, FL was applied to realize ultra reliable and low latency Vehicle-to-Vehicle (V2V) communications while protecting the sensitive data (i.e., queue state information) of users [254]. Specially, in IRS-assisted communication system, the CSI between user and IRS is actually a category of private data, which closely related to the location information of users.

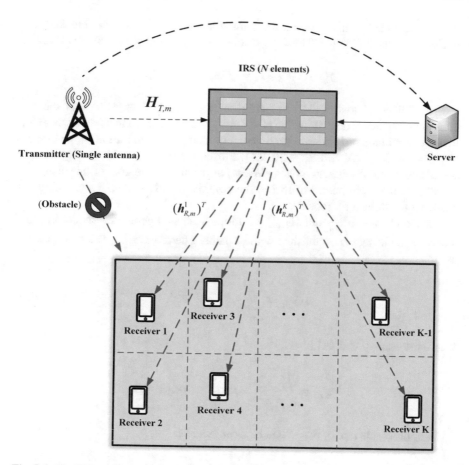

Fig. 9.4 The IRS-assisted wireless communication system

9.2.2 Problem Formulation

As shown in Fig. 9.4, a transmitter is considered to communicate with K receivers assisted by IRS, meanwhile, a server is connected to IRS for data processing. And we assume transmitter and receivers are equipped with single antenna, respectively. The IRS has N reflecting elements. It is worth noting that the transmitter can be the base station, access point (AP), or user's device.

The channel model is based on Orthogonal Frequency Division Multiplexing (OFDM) with M subcarriers, the channels from transmitter and the kth receiver to IRS are defined as $\mathbf{H}_{T,m} \in \mathbb{C}^{N \times 1}$ and $\mathbf{h}_{R,m}^k \in \mathbb{C}^{N \times 1}$, respectively, where $m = 1, 2, \ldots, M$, $k = 1, 2, \ldots, K$. And x_m^k is denoted as the transmitting signal from transmitter to the kth receiver over the m_{th} subcarrier with the power $\left\| x_m^k \right\|^2 = \frac{P}{M}$, in which P represents the total transmit power of each link. In particular, the direct line-of-sight (LOS) link between the transmitter and the receiver is blocked (i.e., occlusion of the buildings). Thus, the received signal at the kth receiver is given as

$$y_m^k = ((\mathbf{h}_{R,m}^k)^T \mathbf{\Phi}_m^k \mathbf{H}_{T,m}) x_m^k + \omega_m^k, \tag{9.25}$$

where a diagonal matrix $\mathbf{\Phi}_m^k = diag[\psi_1, \psi_2, \dots, \psi_N] \in \mathbb{C}^{N \times N}$ is the IRS configuration matrix describing the phase shift effect of IRS on incident signal. Note that the amplitude of incident signal does not change, which means $\psi_n = e^{j\theta_n}$ for any $n = 1, 2, \dots, N$ and $\theta_n \in [0, 2\pi]$. Furthermore, $\omega_m^k \sim \mathcal{CN}(0, \sigma_m^2)$ denotes the additive white Gaussian noise (AWGN) at receivers. The goal of this chapter is to improve the performance of IRS in the communication system by adjusting $\mathbf{\Phi}_m^k$ based on the trained FL model.

For the channel $\mathbf{h}_{R,m}^k$ and $\mathbf{H}_{T,m}$, we adopt wideband geometric model in [254] where each channel is established with L paths. Therefore, $\mathbf{h}_{R,m}^k$ can be expressed as

$$\mathbf{h}_{R,m}^k = \sum_{d=0}^{D-1} \mathbf{h}_{R,d}^k e^{-j\frac{2\pi m}{M}d}, \tag{9.26}$$

where the delay-d channel is

$$\mathbf{h}_{R,d}^k = \sqrt{\frac{N}{\rho}} \sum_{l=1}^{L} \gamma_l p(dT - \eta) \mathbf{a}(\theta_l, \phi_l), \tag{9.27}$$

The achievable rate of user k can be expressed as

$$R^k = \frac{1}{M} \sum_{m=1}^{M} \log_2 \left(1 + r \left| (\mathbf{h}_{R,m}^k)^T \mathbf{\Phi}_m^k \mathbf{H}_{T,m} \right|^2 \right), \tag{9.28}$$

where $r = \frac{P}{M\sigma_m^2}$ denotes the signal to noise ratio (SNR). For the sake of simplicity, we assume that there is no difference for the IRS configuration matrix between each subcarriers of the same user, which means $\mathbf{\Phi}_1^k =, \dots, = \mathbf{\Phi}_m^k = \mathbf{\Phi}^k$. Consequently, a predefined set $\mathbf{O}$ can be built which consists of a lot of predefined configuration matrix.

Obviously, our objective is to train the DNN model to establish the mapping function between $\mathbf{H}_{T,m}, \mathbf{h}_{R,m}^k$ and the optimal IRS configuration $\widehat{\mathbf{\Phi}}$ by searching over $\mathbf{O}$ to further realize the rate maximization.

The searching process can be described as

$$\widehat{\mathbf{\Phi}} = \arg\max_{\mathbf{\Phi}^k \in \mathbf{O}} \sum_{m=1}^{M} \log_2 \left(1 + r \left| (\mathbf{h}_{R,m}^k)^T \mathbf{\Phi}^k \mathbf{H}_{T,m} \right|^2 \right), \tag{9.29}$$

so the optimal average achievable rate $\widehat{R}^k$ at receiver k can obtained as

$$\widehat{R}^k = \frac{1}{M}\sum_{m=1}^{M}\log_2\left(1 + r\left|(\mathbf{h}_{R,m}^k)^T\widehat{\boldsymbol{\Phi}}\mathbf{H}_{T,m}\right|^2\right). \tag{9.30}$$

There is a difficulty that such a huge amount of data about perfect CSI with massive IRS elements will increase the training burden in the implementation of the above scheme. The traditional method of CSI acquisition is to connect the RF receive chain with all elements of IRS, which is extremely expensive and complex to carry out [255]. Thus, we adopt the architecture of IRS with sparse sensors [254]. A tiny fraction of IRS' elements $\overline{N} \ll N$ are active elements with extra ability about channel sensing, which means it can switch the working mode from normal reflective elements to sensor mode when channel estimation is performed. And these active elements are randomly distributed among IRS elements.

9.2.3 FL Assisted Optimal Beam Reflection

FL has attracted more and more interests as a branch of distributed learning with several irreplaceable advantages such as privacy protection and distributed computation. Some work about performance optimization of FL over wireless networks has been done and we assume the wireless links for FL are stable [29]. Here we introduce a novel scheme that combines FL with IRS in communication system. The basic structure of FL is illustrated in Fig. 9.5. Each receiver $U^k (\forall k = 1, 2, \ldots, K)$ participating in training process has its unique dataset S^k, which means it is not accessible for others, processes on local device merely and we assume the same size for all these datasets.

Specifically, the standard federated learning algorithm is adopted (i.e., Federated Averaging). Thus, the whole process can be summarized as three steps:

- Training the local model $\mathbf{W}_i^k$ according to the local dataset S^k on local device U^k, which refers to the receiver in Sect. 9.2.1, where $\mathbf{W}_i^k$ represents the local model trained by the kth device after the i_{th} training. It is worth mentioning that the input and output of the local model are the sampled channel vector and the corresponding rate vector discussed later.
- Aggregating all local models $\mathbf{W}_i^1, \mathbf{W}_i^2, \ldots, \mathbf{W}_i^K$ at central server to generate a global model $\mathbf{W}_{i+1}$. This process of aggregation can be expressed as

$$\mathbf{W}_{i+1} = \frac{1}{K}\sum_{k=1}^{K}\mathbf{W}_i^k. \tag{9.31}$$

- Downloading the global model $\mathbf{W}_{i+1}$ to each device as the initial configuration for next training round.

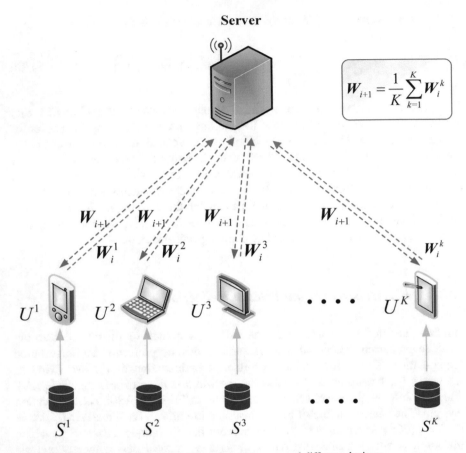

Fig. 9.5 The framework of Federated Learning among several different devices

The optimal model $\mathbf{W}_I$ can be obtained by repeating these procedures until the model converges where I denotes the total training times. The basic of FL is DL, which can be separated into training and validation. For the training phase, the first step is the dataset construction. And the primary task of dataset construction is CSI acquisition based on channel estimation through active elements of IRS. For the simplicity, we assume the channel $\mathbf{H}_{T,m}$ remains constant, so the channel estimation is only for $\mathbf{h}_{R,m}^k$. In order to ensure the facticity of CSI, a random receiving noise will be added to the estimated channel which can be written as

$$\widetilde{\mathbf{h}}_{R,m}^k = \overline{\mathbf{h}}_{R,m}^k + n_m^k, \tag{9.32}$$

Algorithm 17 Optimal beam reflection based on federated learning (OBR-FL)

1: **Learning Phase:**
2: **for** device/receiver $k = 1, 2, \ldots, K$ **do**
3:     **for** $t = 1, 2, \ldots, \xi$ **do**
4:         Sampling the channel information;
5:         $\widetilde{\mathbf{h}}^{k_t} = V\left(\left[\widetilde{\mathbf{h}}^{k_t}_{R,1}, \widetilde{\mathbf{h}}^{k_t}_{R,2}, \ldots, \widetilde{\mathbf{h}}^{k_t}_{R,M}\right]\right);$
6:         Scanning $\mathbf{O}$ and receive a rate set $\mathbf{r}^{k_t}$;
7:         Select out the optimal rate $\widehat{R}^{k_t}$;
8:         Construct data points $(\widetilde{\mathbf{h}}^{k_t}, \widehat{R}^{k_t})$ and add it into local dataset S^k;
9:     **end for**
10:     Local dataset construction finished;
11:     $S^k = \left[(\widetilde{\mathbf{h}}^{k_1}, \widehat{R}^{k_1}), (\widetilde{\mathbf{h}}^{k_2}, \widehat{R}^{k_2}), \ldots, (\widetilde{\mathbf{h}}^{k_\xi}, \widehat{R}^{k_\xi})\right];$
12:     **for** $i = 1, 2, \ldots, I$ **do**
13:         Train DNN at local device, generate local model $\mathbf{W}^k_i$;
14:         Aggregate local models to produce the global model $\mathbf{W}_{i+1}$;
15:         Download $\mathbf{W}_{i+1}$ to each device as initial model configuration.
16:     **end for**
17:     Optimal model $\mathbf{W}_I$ is obtained.
18: **end for**
19: **Validation Phase:**
20: Channel $\widetilde{\mathbf{h}}^k$ sampling;
21: Prediction with the model $\mathbf{W}_I$;
22: Optimal rate $\widehat{R}^{k_t}$ selection.

---

where $\overline{\mathbf{h}}^k_{R,m}$ is the sampled channel. Then we build the vector

$$\widetilde{\mathbf{h}}^k = V\left(\left[\widetilde{\mathbf{h}}^k_{R,1}, \widetilde{\mathbf{h}}^k_{R,2}, \ldots, \widetilde{\mathbf{h}}^k_{R,M}\right]\right)$$

that contains all subcarriers' CSI, where V denotes vector.

Based on the supervised learning, label matching is followed which takes two main parts:

- Establishing a rate vector $\mathbf{r}^k = \left[R^k_1, R^k_2, \ldots, R^k_{|\mathbf{O}^k|}\right]$ by scanning the predefined configuration set $\mathbf{O}$ and each $\mathbf{\Phi}^k$ in $\mathbf{O}$ is applied to formula (9.28) in order.
- Selecting the highest rate $\widehat{R}^k$ among $\mathbf{r}^k$ as the corresponding label.

After the CSI acquisition and label matching, the data point $(\widetilde{\mathbf{h}}^k, \widehat{R}^k)$ can be added to the dataset S^k. It is worth noting that, as shown in Fig. 9.6, the local dataset of each user device consists of its own historical data, which includes the historical CSI and the optimal rate that recorded by itself when the device in different locations of the region. The size of local dataset is denoted by ξ so that the historical data points can be written as $(\widetilde{\mathbf{h}}^{k_t}, \widehat{R}^{k_t})$ for $\forall t = 1, 2, \ldots, \xi$. Consequently, the federated dataset $S^k = \left[(\widetilde{\mathbf{h}}^{k_1}, \widehat{R}^{k_1}), (\widetilde{\mathbf{h}}^{k_2}, \widehat{R}^{k_2}), \ldots, (\widetilde{\mathbf{h}}^{k_\xi}, \widehat{R}^{k_\xi})\right]$ is constructed by utilizing the previous data generation approach.

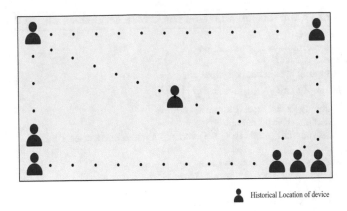

Historical Location of device

Fig. 9.6 The local dataset of each device that consists of historical location information

We adopt the forward verification about trained model with the validation set. The process can be summarized as channel sampling, candidate rate vector generating, and optimal rate selecting. In this chapter, the Multi-Layer-Perceptron (MLP) is adopted as the basic DNN architecture which includes 6 full-connected layers. Rectified Linear Units (ReLU) and Root-Mean-Squared-Error (RMSE) are selected as active function and loss function, respectively. Meanwhile, we employ Stochastic Gradient Descent (SGD) for gradient descent. The overall algorithm is presented in Algorithm 1.

9.2.4 Simulation

The experimental scenario we build is illustrated in Fig. 9.7. BS 7 is activated as IRS and row R1850 column 90 is the location where the transmitter fixed. Meanwhile, receivers grid is constructed with 65,160 points from R2001 to R2360 where each row contains 181 points. We build local datasets by dividing the region of grid to 6 parts which means $K = 6$ and each independent area is consisted of 60 rows with $\xi = 10{,}860$ points totally, while 80% and 20% of these points are training set and validation set, respectively. The default configuration about IRS is 24×24 ($N = 576$) elements working at 28 GHz operating frequency based on 100 MHz OFDM channel with $M = 512$ subcarriers. However, in order to reduce the complexity of DNN, only the first $M_{FL} = 64$ subcarriers are selected to construct local dataset. For transmitter and receivers, all of them are equipped with single antenna that have 5dBi gain. And we implement Discrete Fourier transform (DFT) to establish the IRS configuration matrix set **O**. Meanwhile, the same structure of DNN mentioned previously is adopted to the centralized ML for comparison.

The convergence trend of FL and ML based algorithm is shown in Fig. 9.8, which indicates the rationality of the proposed algorithm. It is obviously demonstrated

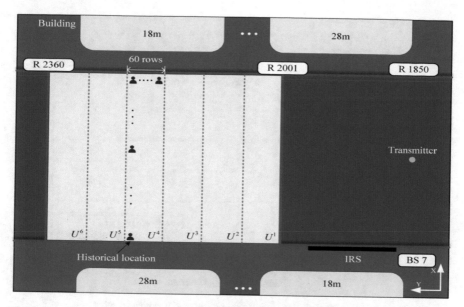

Fig. 9.7 Experimental scenario for FL-based IRS system

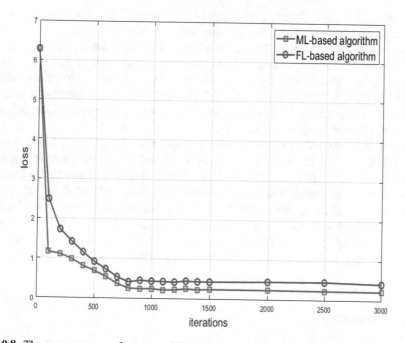

Fig. 9.8 The convergence performance of FL-based algorithm and ML-based algorithm

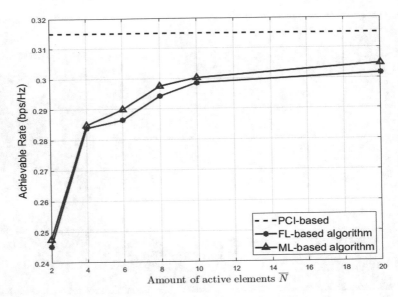

Fig. 9.9 The achievable rate based on federated learning and machine learning

that the value of loss function tends to be stable after 800 times iterations and
the proposed algorithm based on FL can almost achieve the similar convergence
performance compared with ML with a little bit lower convergence speed. For
the difference of convergence speed, it is mainly caused by the time delay during
the process of model updating in FL. Meanwhile, it is known that the effect of
centralized ML is better than distributed ML under the same amount of training
data and training time, which explain the distinction of convergence effect between
proposed algorithm and centralized ML.

Figure 9.9 shows the achievable rate performance of different schemes versus
the number of active elements $\overline{N} = 2, 4, 6, 8, 10$, and 20. The result is generated at
frequency 28 GHz and $L = 10$ paths. The achievable rate increases as the increase
of $\overline{N}$. It is worth noting that the achievable rate up to 90% of ideal value with
$\overline{N} = 8$ which demonstrate that only relatively few active elements are required
for the proposed algorithm to achieve near-optimal rate performance. Furthermore,
as shown in Fig. 9.9 the achievable rate performance of FL-based algorithm can
effectively approach to that of the ML-based algorithm.

9.3 Summary

In this chapter, we have presented two applications, such smart industries and
IRS that can be efficiently enabled via federated learning. For smart industries,
we proposed dispersed federated learning for smart industries to offer robust and

resource efficient learning. An optimization problem is formulated to minimize packet error rate and transmission energy by optimizing resource allocation and transmit power allocation. We applied BSUM-based solution due to the non-convex nature of the formulated problem. In the second part, we introduce an approach based on DL to design the IRS configuration matrix (i.e., the phase shift coefficient matrix), which utilized the sampled CSI for IRS training. In this way, the achievable rate at the receiver can be maximized.

References

1. C. Zhang, P. Patras, H. Haddadi, Deep learning in mobile and wireless networking: a survey. IEEE Commun. Surv. Tutorials **21**(3), 2224–2287 (2019)
2. C.V.N. Index, Forecast and methodology, 2016–2021. *White Paper, Cisco Public*, vol. 6, (2017)
3. W. Saad, M. Bennis, M. Chen, A vision of 6g wireless systems: applications, trends, technologies, and open research problems. IEEE Netw. **34**(3), 134–142 (2019)
4. L.U. Khan, I. Yaqoob, M. Imran, Z. Han, C.S. Hong, 6g wireless systems: a vision, architectural elements, and future directions. IEEE Access **8**, 147029–147044 (2020)
5. P. Yang, Y. Xiao, M. Xiao, S. Li, 6g wireless communications: vision and potential techniques. IEEE Netw. **33**(4), 70–75 (2019)
6. L.U. Khan, W. Saad, D. Niyato, Z. Han, C.S. Hong, Digital-twin-enabled 6g: vision, architectural trends, and future directions (2021). arXiv preprint arXiv:2102.12169
7. S. Ali, W. Saad, N. Rajatheva, K. Chang, D. Steinbach, B. Sliwa, C. Wietfeld, K. Mei, H. Shiri, H.-J. Zepernick et al., 6g white paper on machine learning in wireless communication networks (2020). arXiv preprint arXiv:2004.13875
8. M. Munir, N. H. Tran, W. Saad, C.S. Hong et al., Multi-agent meta-reinforcement learning for self-powered and sustainable edge computing systems (2020). arXiv preprint arXiv:2002.08567
9. M. Alsenwi, N.H. Tran, M. Bennis, S.R. Pandey, A.K. Bairagi, C.S. Hong, Intelligent resource slicing for eMBB and URLLC coexistence in 5g and beyond: a deep reinforcement learning based approach (2020). arXiv preprint arXiv:2003.07651
10. A. Ndikumana, N.H. Tran, K.T. Kim, C.S. Hong et al., Deep learning based caching for self-driving cars in multi-access edge computing. IEEE Trans. Intell. Transp. Syst. **22**(5), 2862–2877 (2021)
11. H. Khan, A. Elgabli, S. Samarakoon, M. Bennis, C.S. Hong, Reinforcement learning-based vehicle-cell association algorithm for highly mobile millimeter wave communication. IEEE Trans. Cogn. Commun. Netw. **5**(4), 1073–1085 (2019)
12. K. Thar, T.Z. Oo, Y.K. Tun, K.T. Kim, C.S. Hong et al., A deep learning model generation framework for virtualized multi-access edge cache management. IEEE Access **7**, 62734–62749 (2019)
13. D. Chen, Y.-C. Liu, B. Kim, J. Xie, C.S. Hong, Z. Han, Edge computing resources reservation in vehicular networks: a meta-learning approach. IEEE Trans. Veh. Tech. **69**(5), 5634–5646 (2020)
14. I. Raicu, I. Foster, A. Szalay, G. Turcu, Astroportal: a science gateway for large-scale astronomy data analysis, in *Teragrid Conference* (2006), pp. 12–15

15. L.U. Khan, W. Saad, Z. Han, E. Hossain, C.S. Hong, Federated learning for internet of things: recent advances, taxonomy, and open challenges (2020). arXiv preprint arXiv:2009.13012

16. W.Y.B. Lim, N.C. Luong, D.T. Hoang, Y. Jiao, Y.-C. Liang, Q. Yang, D. Niyato, C. Miao, Federated learning in mobile edge networks: a comprehensive survey. IEEE Commun. Surv. Tutorials **22**(3), 2031–2063 (2020)

17. J. Verbraeken, M. Wolting, J. Katzy, J. Kloppenburg, T. Verbelen, J.S. Rellermeyer, A survey on distributed machine learning. ACM Comput. Surv. **53**(2), 1–33 (2020)

18. B. McMahan, E. Moore, D. Ramage, S. Hampson, B.A.Y. Arcas, Communication-efficient learning of deep networks from decentralized data, in *Artificial Intelligence and Statistics.* PMLR (2017), pp. 1273–1282

19. L.U. Khan, S.R. Pandey, N.H. Tran, W. Saad, Z. Han, M.N. Nguyen, C.S. Hong, Federated learning for edge networks: resource optimization and incentive mechanism. IEEE Commun. Mag. **58**(10), 88–93 (2020)

20. L.U. Khan, W. Saad, Z. Han, C.S. Hong, Dispersed federated learning: vision, taxonomy, and future directions (2020). arXiv preprint arXiv:2008.05189

21. R.C. Geyer, T. Klein, M. Nabi, Differentially private federated learning: a client level perspective (2017). arXiv preprint arXiv:1712.07557

22. K. Wei, J. Li, M. Ding, C. Ma, H.H. Yang, F. Farokhi, S. Jin, T.Q. Quek, H.V. Poor, Federated learning with differential privacy: algorithms and performance analysis. IEEE Trans. Inf. Forensics Secur. **15**, 3454–3469 (2020)

23. M. Seif, R. Tandon, M. Li, Wireless federated learning with local differential privacy, in *2020 IEEE International Symposium on Information Theory (ISIT)* (IEEE, Piscataway, 2020), pp. 2604–2609

24. Z. Ji, Z.C. Lipton, C. Elkan, Differential privacy and machine learning: a survey and review (2014). arXiv preprint arXiv:1412.7584

25. Understanding differential privacy. [Online; Accessed 11 Mar 2021]. https:// towardsdatascience.com/understanding-differential-privacy-85ce191e198a

26. S.R. Pandey, N.H. Tran, M. Bennis, Y.K. Tun, A. Manzoor, C.S. Hong, A crowdsourcing framework for on-device federated learning. IEEE Trans. Wirel. Commun. **19**(5), 3241–3256 (2020)

27. T.H. T. Le, N.H. Tran, Y.K. Tun, Z. Han, C.S. Hong, Auction based incentive design for efficient federated learning in cellular wireless networks, in *2020 IEEE Wireless Communications and Networking Conference (WCNC)* (IEEE, Piscataway, 2020), pp. 1–6

28. J. Kang, Z. Xiong, D. Niyato, S. Xie, J. Zhang, Incentive mechanism for reliable federated learning: a joint optimization approach to combining reputation and contract theory. IEEE Internet Things J. **6**(6), 10700–10714 (2019)

29. M. Chen, Z. Yang, W. Saad, C. Yin, H.V. Poor, S. Cui, A joint learning and communications framework for federated learning over wireless networks. IEEE Trans. Wirel. Commun. **20**(1), 269–283 (2020)

30. L.U. Khan, M. Alsenwi, Z. Han, C.S. Hong, Self organizing federated learning over wireless networks: a socially aware clustering approach, in *2020 International Conference on Information Networking (ICOIN)* (IEEE, Piscataway, 2020), pp. 453–458

31. N.H. Tran, W. Bao, A. Zomaya, M.N. Nguyen, C.S. Hong, Federated learning over wireless networks: optimization model design and analysis, in *IEEE INFOCOM 2019-IEEE Conference on Computer Communications* (IEEE, Piscataway, 2019), pp. 1387–1395

32. [Online; Accessed 2 Mar 2021]. https://labelyourdata.com/articles/history-of-machine-learning-how-did-it-all-start/

33. W.S. McCulloch, W. Pitts, A logical calculus of the ideas immanent in nervous activity. Bull. Math. Biophys. **5**(4), 115–133 (1943)

34. A.M. Turing, Computing machinery and intelligence, in *Parsing the Turing Test* (Springer, Berlin, 2009), pp. 23–65

35. C. Machinery, Computing machinery and intelligence A.M. Turing. Mind **59**(236), 433 (1950)

36. [Online; Accessed 2 Mar 2021]. http://www-formal.stanford.edu/jmc/slides/dartmouth/dartmouth/node1.html
37. [Online; Accessed 2 Mar 2021]. https://builtin.com/artificial-intelligence/deep-learning-history
38. T. Cover, P. Hart, Nearest neighbor pattern classification. IEEE Trans. Inf. Theory **13**(1), 21–27 (1967)
39. K. Fukushima, S. Miyake, Neocognitron: a self-organizing neural network model for a mechanism of visual pattern recognition, in *Competition and Cooperation in Neural Nets* (Springer, Berlin, 1982), pp. 267–285
40. T.J. Sejnowski, C.R. Rosenberg, Nettalk: a parallel network that learns to read aloud. johns hopkins university electrical engineering and computer science technical report. EEC **86**(1) (1986)
41. N. McCulloch, M. Bedworth, J. Bridle, Netspeak—a re-implementation of nettalk. Comput. Speech Lang. **2**(3–4), 289–302 (1987)
42. R.E. Schapire, The strength of weak learnability. Mach. Learn. **5**(2), 197–227 (1990)
43. T.K. Ho, Random decision forests, in *Proceedings of 3rd International Conference on Document Analysis and Recognition*, vol. 1 (IEEE, Piscataway, 1995), pp. 278–282
44. [Online; Accessed 2 Mar 2021]. https://towardsdatascience.com/the-deep-history-of-deep-learning-3bebeb810fb2
45. G.E. Hinton, Deep belief networks. Scholarpedia **4**(5), 5947 (2009)
46. [Online; Accessed 2 Mar 2021]. https://www.doc.ic.ac.uk/~jce317/history-machine-learning.html
47. P.A. Bernstein, E. Newcomer, *Principles of Transaction Processing* (Morgan Kaufmann, Los Altos, 2009)
48. D. Alistarh, Distributed machine learning: a brief overview. [Online; Accessed March, 2020]. [Online]. https://www.podc.org/data/podc2018/podc2018-tutorial-alistarh.pdf
49. J. Qiu, Q. Wu, G. Ding, Y. Xu, S. Feng, A survey of machine learning for big data processing. EURASIP J. Adv. Signal Process. **2016**(1), 67 (2016)
50. D. Peteiro-Barral, B. Guijarro-Berdiñas, A survey of methods for distributed machine learning. Progr. Artif. Intell. **2**(1), 1–11 (2013)
51. P. Kairouz, H.B. McMahan, B. Avent, A. Bellet, M. Bennis, A.N. Bhagoji, K. Bonawitz, Z. Charles, G. Cormode, R. Cummings et al., Advances and open problems in federated learning (2019). arXiv preprint arXiv:1912.04977
52. T. Li, M. Sanjabi, A. Beirami, V. Smith, Fair resource allocation in federated learning (2019). arXiv preprint arXiv:1905.10497
53. S. Wang, T. Tuor, T. Salonidis, K.K. Leung, C. Makaya, T. He, K. Chan, Adaptive federated learning in resource constrained edge computing systems. IEEE J. Sel. Areas Commun. **37**(6), 1205–1221 (2019)
54. V. Smith, C.-K. Chiang, M. Sanjabi, A. Talwalkar, Federated multi-task learning (2017). arXiv preprint arXiv:1705.10467
55. M. Jaggi, V. Smith, M. Takáč, J. Terhorst, S. Krishnan, T. Hofmann, M.I. Jordan, Communication-efficient distributed dual coordinate ascent (2014). arXiv preprint arXiv:1409.1458
56. Z. Han, D. Niyato, W. Saad, T. Başar, A. Hjørungnes, Game theory in wireless and communication networks: theory, models, and applications (Cambridge University Press, Cambridge, 2012)
57. T. Li, A.K. Sahu, M. Zaheer, M. Sanjabi, A. Talwalkar, V. Smith, Federated optimization in heterogeneous networks (2018). arXiv preprint arXiv:1812.06127
58. J. Konečný, H.B. McMahan, D. Ramage, P. Richtárik, Federated optimization: distributed machine learning for on-device intelligence (2016). arXiv:1610.02527 [cs]
59. C. Ma, J. Konečný, M. Jaggi, V. Smith, M.I. Jordan, P. Richtárik, M. Takáč, Distributed optimization with arbitrary local solvers. Optim. Methods Softw. **32**(4), 813–848 (2017)
60. J. Konečný, Z. Qu, P. Richtárik, Semi-stochastic coordinate descent. Optim. Methods Softw. **32**(5), 993–1005 (2017)

61. A.P. Miettinen, J.K. Nurminen, Energy efficiency of mobile clients in cloud computing, in *USENIX HotCloud'10*, Berkeley, CA, USA (2010), p. 4
62. T.D. Burd, R.W. Brodersen, Processor design for portable systems. J. VLSI Signal Process. Syst. **13**(2–3), 203–221 (1996)
63. S. Kandukuri, S. Boyd, Optimal power control in interference-limited fading wireless channels with outage-probability specifications. IEEE Trans. Wirel. Commun. **1**(1), 46–55 (2002)
64. B. Prabhakar, E.U. Biyikoglu, A.E. Gamal, Energy-efficient transmission over a wireless link via lazy packet scheduling, in *IEEE INFOCOM 2001*, vol. 1 (2001), pp. 386–394
65. A. Wächter, L.T. Biegler, On the implementation of an interior-point filter line-search algorithm for large-scale nonlinear programming. Math. Program. **106**(1), 25–57 (2006)
66. M. Chen, U. Challita, W. Saad, C. Yin, M. Debbah, Artificial neural networks-based machine learning for wireless networks: a tutorial. IEEE Commun. Surveys Tut. **21**(4), 3039–3071 (2019)
67. K. Bonawitz, H. Eichner, W. Grieskamp, D. Huba, A. Ingerman, V. Ivanov, C.M. Kiddon, J. Konecny, S. Mazzocchi, B. McMahan, T.V. Overveldt, D. Petrou, D. Ramage, J. Roselander, Towards federated learning at scale: system design, in *Proc. Systems and Machine Learning Conference*, Stanford, CA, USA, 2019
68. X. Wang, Y. Han, C. Wang, Q. Zhao, X. Chen, M. Chen, In-edge AI: intelligentizing mobile edge computing, caching and communication by federated learning. IEEE Netw. **33**(5), 156–165 (2019)
69. E. Jeong, S. Oh, J. Park, H. Kim, M. Bennis, S.L. Kim, Multi-hop federated private data augmentation with sample compression, in *Proc. International Joint Conference on Artificial Intelligence Workshop on Federated Machine Learning for User Privacy and Data Confidentiality*, Macao, China, 2019
70. Z. Yang, M. Chen, W. Saad, C.S. Hong, M. Shikh-Bahaei, Energy efficient federated learning over wireless communication networks. IEEE Trans. Wirel. Commun. **20**(3), 1935–1949 (2021)
71. T. Li, A.K. Sahu, A. Talwalkar, V. Smith, Federated learning: challenges, methods, and future directions. IEEE Signal Process. Mag. **37**(3), 50–60 (2020)
72. J. Konečný, H.B. McMahan, F.X. Yu, P. Richtárik, A.T. Suresh, D. Bacon, Federated learning: strategies for improving communication efficiency (2016). arXiv preprint arXiv:1610.05492
73. M. Chen, O. Semiari, W. Saad, X. Liu, C. Yin, Federated echo state learning for minimizing breaks in presence in wireless virtual reality networks. IEEE Trans. Wirel. Commun. **19**(1), 177–191 (2020)
74. J. Konečný, B. McMahan, D. Ramage, Federated optimization: distributed optimization beyond the datacenter (2015). arXiv preprint arXiv:1511.03575
75. S. Samarakoon, M. Bennis, W. Saad, M. Debbah, Distributed federated learning for ultra-reliable low-latency vehicular communications. IEEE Trans. Commun. **68**(2), 1146–1159 (2020)
76. S. Ha, J. Zhang, O. Simeone, J. Kang, Coded federated computing in wireless networks with straggling devices and imperfect CSI, in *Proc. IEEE Int. Symp. Inf. Theory (ISIT)*, Paris, France, 2019
77. J. Park, S. Samarakoon, M. Bennis, M. Debbah, Wireless network intelligence at the edge. Proc. IEEE **107**(11), 2204–2239 (2019)
78. Q. Zeng, Y. Du, K. Huang, K.K. Leung, Energy-efficient radio resource allocation for federated edge learning, in *Proc. IEEE Int. Conf. Commun. Workshop*, Dublin, Ireland, 2020
79. Z. Zhao, C. Feng, H.H. Yang, X. Luo, Federated-learning-enabled intelligent fog radio access networks: fundamental theory, key techniques, and future trends. IEEE Wirel. Commun. **27**(2), 22–28 (2020)
80. T.T. Vu, D.T. Ngo, N.H. Tran, H.Q. Ngo, M.N. Dao, R.H. Middleton, Cell-free massive MIMO for wireless federated learning. IEEE Trans. Wirel. Commun. **19**(10), 6377–6392 (2020)

81. M. Chen, H.V. Poor, W. Saad, S. Cui, Convergence time optimization for federated learning over wireless networks. IEEE Trans. Wirel. Commun. **20**(4), 2457–2471 (2021)

82. M. Chen, N. Shlezinger, H.V. Poor, Y.C. Eldar, S. Cui, Communication-efficient federated learning. Proc. Nat. Acad. Sci. **118**(17), e2024789118 (2021)

83. H.H. Yang, Z. Liu, T.Q.S. Quek, H.V. Poor, Scheduling policies for federated learning in wireless networks. IEEE Tans. Commun. **68**(1), 317–333 (2020)

84. Y. Xi, A. Burr, J. Wei, D. Grace, A general upper bound to evaluate packet error rate over quasi-static fading channels. IEEE Tans. Wirel. Commun. **10**(5), 1373–1377 (2011)

85. Y. Pan, C. Pan, Z. Yang, M. Chen, Resource allocation for D2D communications underlaying a NOMA-based cellular network. IEEE Wirel. Commun. Lett. **7**(1), 130–133 (2018)

86. M.P. Friedlander, M. Schmidt, Hybrid deterministic-stochastic methods for data fitting. SIAM J. Sci. Comput. **34**(3), A1380–A1405 (2012)

87. R. Jonker, T. Volgenant, Improving the Hungarian assignment algorithm. Oper. Res. Lett. **5**(4), 171–175 (1986)

88. Y. LeCun, The MNIST database of handwritten digits. http://yann.lecun.com/exdb/mnist/

89. M. Chen, H.V. Poor, W. Saad, S. Cui, Wireless communications for collaborative federated learning. IEEE Commun. Mag. **58**(12), 48–54 (2020)

90. J. Kang, Z. Xiong, D. Niyato, Y. Zou, Y. Zhang, M. Guizani, Reliable federated learning for mobile networks. IEEE Wirel. Commun. **27**(2), 72–80 (2020)

91. S. Wang, M. Chen, C. Yin, W. Saad, C.S. Hong, S. Cui, H.V. Poor, Federated learning for task and resource allocation in wireless high altitude balloon networks. IEEE Internet Things J. (2021)

92. Y. Wang, Y. Yang, T. Luo, Federated convolutional auto-encoder for optimal deployment of UAVs with visible light communications, in *Proc. IEEE International Conference on Communications Workshops (ICC Workshops)*, Dublin, Ireland

93. T. Zeng, O. Semiari, M. Chen, W. Saad, M. Bennis, Federated learning on the road: autonomous controller design for connected and autonomous vehicles (2021). arXiv preprint arXiv:2102.03401

94. M. Chen, D. Gündüz, K. Huang, W. Saad, M. Bennis, A.V. Feljan, H.V. Poor, Distributed learning in wireless networks: recent progress and future challenges (2021). arXiv preprint arXiv:2104.02151

95. C. Liu, C. Guo, Y. Yang, M. Chen, H.V. Poor, S. Cui, Optimization of user selection and bandwidth allocation for federated learning in VLC/RF systems, in *Proc. IEEE Wireless Communications and Networking Conference (WCNC)*, Nanjing, China, 2021

96. N. Shlezinger, M. Chen, Y.C. Eldar, H.V. Poor, S. Cui, Uveqfed: universal vector quantization for federated learning. IEEE Trans. Signal Process. **69**, 500–514 (2021)

97. S. Niknam, H.S. Dhillon, J.H. Reed, Federated learning for wireless communications: motivation, opportunities, and challenges. IEEE Commun. Mag. **58**(6), 46–51 (2020)

98. G. Zhu, D. Liu, Y. Du, C. You, J. Zhang, K. Huang, Toward an intelligent edge: wireless communication meets machine learning. IEEE Commun. Mag. **58**(1), 19–25 (2020)

99. A. Nedic, A. Olshevsky, M.G. Rabbat, Network topology and communication-computation tradeoffs in decentralised optimization. Proc. IEEE **106**(5), 953–976 (2018)

100. L.G. Jaimes, I.J. Vergara-Laurens, A. Raij, A survey of incentive techniques for mobile crowd sensing. IEEE Internet Things J. **2**(5), 370–380 (2015)

101. S.A. Kazmi, N.H. Tran, T.M. Ho, C.S. Hong, Hierarchical matching game for service selection and resource purchasing in wireless network virtualization. IEEE Commun. Lett. **22**(1), 121–124 (2017)

102. T.H.T. Le, N.H. Tran, T. LeAnh, C.S. Hong, User matching game in virtualized 5g cellular networks, in *2016 18th Asia-Pacific Network Operations and Management Symposium (APNOMS)* (IEEE, Piscataway, 2016), pp. 1–4

103. C. Pham, N.H. Tran, S. Ren, W. Saad, C.S. Hong, Traffic-aware and energy-efficient VNF placement for service chaining: joint sampling and matching approach. IEEE Trans. Serv. Comput. **13**(1), 172–185 (2017)

104. M.N. Nguyen, D. Kim, N.H. Tran, C.S. Hong, Multi-stage Stackelberg game approach for colocation datacenter demand response, in *2017 19th Asia-Pacific Network Operations and Management Symposium (APNOMS)* (IEEE, Piscataway, 2017), pp. 139–144

105. T.M. Ho, N.H. Tran, C.T. Do, S.A. Kazmi, T. LeAnh, C.S. Hong, Data offloading in heterogeneous cellular networks: Stackelberg game based approach, in *2015 17th Asia-Pacific Network Operations and Management Symposium (APNOMS)* (IEEE, Piscataway, 2015), pp. 168–173

106. S. Noreen, N. Saxena, A review on game-theoretic incentive mechanisms for mobile data offloading in heterogeneous networks. IETE Tech. Rev. **34**(sup1), 15–26 (2017)

107. C. Liu, S. Xing, L. Shen, Stackelberg-game based pricing framework for hybrid access control in femtocell network, in *2015 IEEE 26th Annual International Symposium on Personal, Indoor, and Mobile Radio Communications (PIMRC)* (IEEE, Piscataway, 2015), pp. 1153–1156

108. K. Poularakis, G. Iosifidis, L. Tassiulas, A framework for mobile data offloading to leased cache-endowed small cell networks, in *2014 IEEE 11th International Conference on Mobile Ad Hoc and Sensor Systems* (IEEE, Piscataway, 2014), pp. 327–335

109. L. Gao, G. Iosifidis, J. Huang, L. Tassiulas, D. Li, Bargaining-based mobile data offloading. IEEE J. Sel. Areas Commun. **32**(6), 1114–1125 (2014)

110. H. Yu, M.H. Cheung, J. Huang, Cooperative wi-fi deployment: a one-to-many bargaining framework. IEEE Trans. Mobile Comput. **16**(6), 1559–1572 (2016)

111. S. Wang, T. Tuor, T. Salonidis, K.K. Leung, C. Makaya, T. He, K. Chan, When edge meets learning: adaptive control for resource-constrained distributed machine learning (2018). arXiv preprint arXiv:1804.05271

112. H. Kim, J. Park, M. Bennis, S.-L. Kim, "On-device federated learning via blockchain and its latency analysis (2018). arXiv preprint arXiv:1808.03949

113. R.K. Ganti, F. Ye, H. Lei, Mobile crowdsensing: current state and future challenges. IEEE Commun. Mag. **49**(11), 32–39 (2011)

114. S. Boyd, N. Parikh, E. Chu, B. Peleato, J. Eckstein et al., Distributed optimization and statistical learning via the alternating direction method of multipliers. Found. Trends® Mach. Learn. **3**(1), 1–122 (2011)

115. S. Shalev-Shwartz, T. Zhang, Accelerated proximal stochastic dual coordinate ascent for regularized loss minimization, in *International Conference on Machine Learning*, Beijing, China, 2014, pp. 64–72

116. S. Boyd, L. Vandenberghe, *Convex Optimization* (Cambridge University Press, Cambridge, 2004)

117. D. Niyato, M.A. Alsheikh, P. Wang, D.I. Kim, Z. Han, Market model and optimal pricing scheme of big data and internet of things (IoT), in *IEEE International Conference on Communications (ICC)*, Kuala Lumpur, Malaysia, May 2016, pp. 1–6

118. F.N. Iandola, M.W. Moskewicz, K. Ashraf, K. Keutzer, Firecaffe: near-linear acceleration of deep neural network training on compute clusters, in *Proc. of the IEEE Conference on Computer Vision and Pattern Recognition*, Las Vegas, Nevada, June 2016, pp. 2592–2600

119. C. Dinh, N.H. Tran, M.N. Nguyen, C.S. Hong, W. Bao, A. Zomaya, V. Gramoli, Federated learning over wireless networks: convergence analysis and resource allocation (2019). arXiv preprint arXiv:1910.13067

120. Y. Liu, R. Wang, Z. Han, Interference-constrained pricing for d2d networks. IEEE Trans. Wirel. Commun. **16**(1), 475–486 (2017)

121. J. Li, H. Chen, Y. Chen, Z. Lin, B. Vucetic, L. Hanzo, Pricing and resource allocation via game theory for a small-cell video caching system. IEEE J. Sel. Areas Commun. **34**(8), 2115–2129 (2016)

122. B. Faltings, J.J. Li, R. Jurca, Incentive mechanisms for community sensing. IEEE Trans. Comput. **63**(1), 115–128 (2014)

123. Y. Bao, Y. Peng, C. Wu, Z. Li, Online job scheduling in distributed machine learning clusters (2018). arXiv preprint arXiv:1801.00936

124. O. Shamir, N. Srebro, Distributed stochastic optimization and learning, in *52nd Annual Allerton Conference on Communication, Control, and Computing* (IEEE, Piscataway, 2014), pp. 850–857

125. S.D. Conte, C. De Boor, *Elementary Numerical Analysis: An Algorithmic Approach*, vol. 78 (SIAM, 2017)

126. N.C. Luong, P. Wang, D. Niyato, Y.-C. Liang, Z. Han, F. Hou, Applications of economic and pricing models for resource management in 5g wireless networks: a survey. IEEE Commun. Surv. Tutorials **21**(4), 3298–3339 (2018)

127. D. Niyato, N.C. Luong, P. Wang, Z. Han, Auction theory for computer networks (2020)

128. Y. Zhang, C. Lee, D. Niyato, P. Wang, Auction approaches for resource allocation in wireless systems: a survey. IEEE Commun. Surv. Tutorials **15**(3), 1020–1041 (2012)

129. N. Nisan, T. Roughgarden, E. Tardos, V.V. Vazirani, *Algorithmic Game Theory* (Cambridge Univ. Press, Cambridge, 2007)

130. J. Wang, D. Yang, J. Tang, M. C. Gursoy, Enabling radio-as-a-service with truthful auction mechanisms. IEEE Trans. Wirel. Commun. **16**(4), 2340–2349 (2017)

131. P. Klemperer, What really matters in auction design. J. Econ. Perspect. **16**(1), 169–189 (2002)

132. Y. Jiao, P. Wang, D. Niyato, B. Lin, D.I. Kim, Toward an automated auction framework for wireless federated learning services market. IEEE Trans. Mobile Comput. (2020)

133. Y. Hao, Q. Ni, H. Li, S. Hou, Energy-efficient multi-user mobile-edge computation offloading in massive MIMO enabled HetNets, in *IEEE International Conference on Communications (ICC)*, Shanghai, China, 2019, pp. 1–6

134. H.Q. Ngo, E.G. Larsson, T.L. Marzetta, Energy and spectral efficiency of very large multiuser MIMO systems. IEEE Trans. Commun. **61**(4), 1436–1449 (2013)

135. X. Lyu, H. Tian, C. Sengul, P. Zhang, Multiuser joint task offloading and resource optimization in proximate clouds. IEEE Trans. Veh. Technol. **66**(4), 3435–3447 (2016)

136. W. Dinkelbach, On nonlinear fractional programming. Manag. Sci. **13**(7), 492–498 (1967)

137. S. Zaman, D. Grosu, Combinatorial auction-based allocation of virtual machine instances in clouds. J. Parallel Distrib. Comput. **73**(44), 495–508 (2013)

138. A.-L. Jin, W. Song, P. Wang, D. Niyato, P. Ju, Auction mechanisms toward efficient resource sharing for cloudlets in mobile cloud computing. IEEE Trans. Serv. Comput. **9**(6), 895–909 (2015)

139. Data volume of internet of things (IoT) connections worldwide in 2018 and 2025 (in zettabytes). 2020. [Online]. https://www.statista.com/statistics/1017863/worldwide-iot-connecteddevices-data-size/

140. M.I. Jordan, T.M. Mitchell, Machine learning: trends, perspectives, and prospects. Science **349**(6245), 255–260 (2015)

141. D. Chen, C.S. Hong, L. Wang, Y. Zha, Y. Zhang, X. Liu, Z. Han, Matching theory based low-latency scheme for multi-task federated learning in mec networks. IEEE Internet Things J. (2021). https://doi.org/10.1109/JIOT.2021.3053283

142. Q. Yang, Y. Liu, T. Chen, Y. Tong, Federated machine learning: concept and applications. ACM Trans. Intell. Syst. Technol. **10**(2), 1–19 (2019)

143. R. Shokri, M. Stronati, C. Song, V. Shmatikov, Membership inference attacks against machine learning models, in *2017 IEEE Symposium on Security and Privacy (SP)*, San Jose, CA, May 2017, pp. 3–18

144. M. Nasr, R. Shokri, A. Houmansadr, Comprehensive privacy analysis of deep learning, in *2019 ieee symposium on security and privacy*, San Francisco, CA, May 2019

145. M. Abadi, A. Chu, I. Goodfellow, H.B. McMahan, I. Mironov, K. Talwar, L. Zhang, Deep learning with differential privacy, in *Proceedings of the 2016 ACM SIGSAC Conference on Computer and Communications Security*, Seoul, Republic of Korea, October 2016, pp. 308–318

146. K. Bonawitz, V. Ivanov, B. Kreuter, A. Marcedone, H.B. McMahan, S. Patel, D. Ramage, A. Segal, K. Seth, Practical secure aggregation for federated learning on user-held data (2016). arXiv preprint arXiv:1611.04482

147. G. Ács, C. Castelluccia, I have a dream!(differentially private smart metering), in *International Workshop on Information Hiding* (Springer, Berlin, 2011), pp. 118–132
148. S. Hardy, W. Henecka, H. Ivey-Law, R. Nock, G. Patrini, G. Smith, B. Thorne, Private federated learning on vertically partitioned data via entity resolution and additively homomorphic encryption (2017). arXiv preprint arXiv:1711.10677
149. Y. Liu, Y. Kang, C. Xing, T. Chen, Q. Yang, A secure federated transfer learning framework. IEEE Intell. Syst. **35**(4), 70–82 (2020)
150. S. Truex, N. Baracaldo, A. Anwar, T. Steinke, H. Ludwig, R. Zhang, Y. Zhou, A hybrid approach to privacy-preserving federated learning, in *Proceedings of the 12th ACM Workshop on Artificial Intelligence and Security*, Seoul, Republic of Korea, November 2019, pp. 1–11
151. R. Xu, N. Baracaldo, Y. Zhou, A. Anwar, H. Ludwig, Hybridalpha: an efficient approach for privacy-preserving federated learning, in *Proceedings of the 12th ACM Workshop on Artificial Intelligence and Security*, Seoul, Republic of Korea, November 2019, pp. 13–23
152. R.L. Rivest, All-or-nothing encryption and the package transform, in *International Workshop on Fast Software Encryption* (Springer, Berlin, 1997), pp. 210–218
153. D.R. Stinson, Something about all or nothing (transforms). Des. Codes Cryptogr. **22**(2), 133–138 (2001)
154. A. Lewko, T. Okamoto, A. Sahai, K. Takashima, B. Waters, Fully secure functional encryption: attribute-based encryption and (hierarchical) inner product encryption, in *Annual International Conference on the Theory and Applications of Cryptographic Techniques* (Springer, Berlin, 2010), pp. 62–91
155. J. Chotard, E.D. Sans, R. Gay, D.H. Phan, D. Pointcheval, Decentralized multi-client functional encryption for inner product, in *International Conference on the Theory and Application of Cryptology and Information Security* (Springer, Brisbane, 2018), pp. 703–732
156. D. Chen, L.J. Xie, B. Kim, L. Wang, C.S. Hong, L.-C. Wang, Z. Han, Federated learning based mobile edge computing for augmented reality applications, in *International Conference on Computing, Networking and Communications (ICNC)*, Big Island, HI, February 2020
157. H. Zhu, Y. Jin, Multi-objective evolutionary federated learning. IEEE Trans. Neural Netw. Learn. Syst. 1–13 (2019)
158. P. Blanchard, R. Guerraoui, J. Stainer et al., Machine learning with adversaries: byzantine tolerant gradient descent. Adv. Neural Inf. Process. Syst. 119–129 (2017)
159. Y. Chen, L. Su, J. Xu, Distributed statistical machine learning in adversarial settings: byzantine gradient descent. Proc. ACM Measur. Anal. Comput. Syst. **1**(2), 44 (2017)
160. Y. Zhao, M. Li, L. Lai, N. Suda, D. Civin, V. Chandra, Federated learning with non-IID data (2018). arXiv preprint arXiv:1806.00582
161. A. Agarwal, M.J. Wainwright, J.C. Duchi, Distributed dual averaging in networks. Adv. Neural Inf. Process. Syst. 550–558 (2010)
162. A. Kassambara, *Practical Guide to Cluster Analysis in R: Unsupervised Machine Learning*, vol. 1 (STHDA, 2017)
163. D. Arthur, S. Vassilvitskii, k-means++: the advantages of careful seeding, in *Proceedings of the Eighteenth Annual ACM-SIAM Symposium on Discrete Algorithms*, 2007, pp. 1027–1035
164. S.Z. Selim, K. Alsultan, A simulated annealing algorithm for the clustering problem. Pattern Recogn. **24**(10), 1003–1008 (1991)
165. P. Goyal, P. Dollár, R. Girshick, P. Noordhuis, L. Wesolowski, A. Kyrola, A. Tulloch, Y. Jia, K. He, Accurate, large minibatch SGD: training imagenet in 1 hour (2017). arXiv preprint arXiv:1706.02677
166. T. Kohonen, The self-organizing map. Proc. IEEE **78**(9), 1464–1480 (1990)
167. L. Bottou, Y. Bengio, Convergence properties of the k-means algorithms, in *Advances in Neural Information Processing Systems* (1995), pp. 585–592
168. S.Z. Selim, M.A. Ismail, K-means-type algorithms: a generalized convergence theorem and characterization of local optimality. IEEE Trans. Pattern Anal. Mach. Intell. **PAMI-6**(1), 81–87 (1984)
169. F. Hu, Y. Deng, W. Saad, M. Bennis, A.H. Aghvami, Cellular-connected wireless virtual reality: requirements, challenges, and solutions. IEEE Commun. Mag. **58**(5), 105–111 (2020)

170. M. Chen, W. Saad, C. Yin, M. Debbah, Data correlation-aware resource management in wireless virtual reality (VR): an echo state transfer learning approach. IEEE Trans. Commun. **67**(6), 4267–4280 (2019)

171. E. Baştuğ, M. Bennis, M. Médard, M. Debbah, Towards interconnected virtual reality: opportunities, challenges and enablers. IEEE Commun. Mag. **55**(6), 110–117 (2017)

172. M. Chen, W. Saad, C. Yin, Echo-liquid state deep learning for 360° content transmission and caching in wireless VR networks with cellular-connected UAVs. IEEE Trans. Commun. **67**(9), 6386–6400 (2019)

173. X. Ge, L. Pan, Q. Li, G. Mao, S. Tu, Multipath cooperative communications networks for augmented and virtual reality transmission. IEEE Trans. Multimedia **1910**, 2345–2358 (2017)

174. Y. Sun, Z. Chen, M. Tao, H. Liu, Communications, caching and computing for mobile virtual reality: modeling and tradeoff (2018). arXiv preprint arXiv:1806.08928

175. J. Park, M. Bennis, URLLC-eMBB slicing to support VR multimodal perceptions over wireless cellular systems, in *Proc. of IEEE Global Communications Conference (GLOBECOM)*, Abu Dhabi, United Arab Emirates, 2018

176. X. Yang, Z. Chen, K. Li, Y. Sun, N. Liu, W. Xie, Y. Zhao, Communication-constrained mobile edge computing systems for wireless virtual reality: scheduling and tradeoff. IEEE Access **6**, 16665–16677 (2018)

177. X. Liu, Y. Deng, Learning-based prediction, rendering and association optimization for MEC-enabled wireless virtual reality (VR) network. IEEE Trans. Wirel. Commun. (2021)

178. A. Taleb Zadeh Kasgari, W. Saad, M. Debbah, Human-in-the-loop wireless communications: machine learning and brain-aware resource management. IEEE Trans. Commun. **67**(11), 7727–7743 (2019)

179. M.S. Elbamby, C. Perfecto, M. Bennis, K. Doppler, Edge computing meets millimeter-wave enabled VR: paving the way to cutting the cord, in *Proc. of IEEE Wireless Communications and Networking Conference*, Barcelona, Spain, 2018

180. W.C. Lo, C.L. Fan, S.C. Yen, C.H. Hsu, Performance measurements of 360 video streaming to head-mounted displays over live 4G cellular networks, in *Proc. of Asia-Pacific Network Operations and Management Symposium*, Seoul, South Korea, 2017

181. M. Chen, W. Saad, C. Yin, Virtual reality over wireless networks: quality-of-service model and learning-based resource management. IEEE Trans. Commun. **66**(11), 5621–5635 (2018)

182. F. Hu, Y. Deng, A.H. Aghvami, Correlation-aware cooperative multigroup broadcast 360 video delivery network: a hierarchical deep reinforcement learning approach (2020). arXiv preprint arXiv:2010.11347

183. O. Semiari, W. Saad, M. Bennis, M. Debbah, Integrated millimeter wave and sub-6 GHz wireless networks: a roadmap for joint mobile broadband and ultra-reliable low-latency communications. IEEE Wirel. Commun. **26**(2), 109–115 (2019)

184. X. Liu, X. Li, Y. Deng, Learning-based prediction and uplink retransmission for wireless virtual reality (VR) network (2020). arXiv preprint arXiv:2012.12725

185. J. Yin, L. Li, H. Zhang, X. Li, A. Gao, Z. Han, A prediction-based coordination caching scheme for content centric networking, in *Proc. of Wireless and Optical Communication Conference*, Hualien, Taiwan, 2018

186. L. Yao, A. Chen, J. Deng, J. Wang, G. Wu, A cooperative caching scheme based on mobility prediction in vehicular content centric networks. IEEE Trans. Veh. Technol. **67**(6), 5435–5444 (2018)

187. N.T. Nguyen, Y. Wang, H. Li, X. Liu, Z. Han, Extracting typical users' moving patterns using deep learning, in *Proc. of IEEE Global Communications Conference*, Anaheim, CA, USA, 2012

188. M. Chen, M. Mozaffari, W. Saad, C. Yin, M. Debbah, C.S. Hong, Caching in the sky: proactive deployment of cache-enabled unmanned aerial vehicles for optimized quality-of-experience. IEEE J. Sel. Areas Commun. **35**(5), 1046–1061 (2017)

189. O. Esrafilian, R. Gangula, D. Gesbert, Learning to communicate in UAV-aided wireless networks: map-based approaches. IEEE Internet Things J. **6**(2), 1791–1802 (2019)

190. M. Chen, W. Saad, C. Yin, Echo state networks for self-organizing resource allocation in LTE-U with uplink–downlink decoupling. IEEE Trans. Wirel. Commun. **16**(1), 3–16 (2017)

191. O. Semiari, W. Saad, M. Bennis, Joint millimeter wave and microwave resources allocation in cellular networks with dual-mode base stations. IEEE Trans. Wirel. Commun. **16**(7), 4802–4816 (2017)

192. Q. Li, M. Yu, A. Pandharipande, X. Ge, J. Zhang, J. Zhang, Performance of virtual full-duplex relaying on cooperative multi-path relay channels. IEEE Trans. Wirel. Commun. **15**(5), 3628–3642 (2016)

193. Q. Li, M. Yu, A. Pandharipande, X. Ge, Outage analysis of co-operative two-path relay channels. IEEE Trans. Wirel. Commun. **15**(5), 3157–3169 (2016)

194. HTC, HTC Vive. https://www.vive.com/us/

195. Oculus, Mobile VR media overview. https://www.oculus.com/

196. O. Semiari, W. Saad, M. Bennis, Z. Dawy, Inter-operator resource management for millimeter wave multi-hop backhaul networks. IEEE Trans. Wirel. Commun. **16**(8), 5258–5272 (2017)

197. K. Venugopal, M.C. Valenti, R.W. Heath, Device-to-device millimeter wave communications: interference, coverage, rate, and finite topologies. IEEE Trans. Wirel. Commun. **15**(9), 6175–6188 (2016)

198. J. Jerald, *The VR Book: Human-Centered Design for Virtual Reality* (Morgan & Claypool, 2015)

199. J. Chung, H.J. Yoon, H.J. Gardner, Analysis of break in presence during game play using a linear mixed model. ETRI J. **32**(5), 687–694 (2010)

200. S. Scardapane, D. Wang, M. Panella, A decentralized training algorithm for echo state networks in distributed big data applications. Neural Netw. **78**, 65–74 (2016)

201. X. Liu, M. Chen, C. Yin, W. Saad, Analysis of memory capacity for deep echo state networks, in *Proc. of IEEE International Conference on Machine Learning and Applications (ICMLA)*, Orlando, FL, USA, 2018

202. M. Bennis, D. Niyato, A Q-learning based approach to interference avoidance in self-organized femtocell networks, in *Proc. of IEEE Global Communications Conference Workshops*, Miami, FL, USA, 2010

203. A. Hard, K. Rao, R. Mathews, S. Ramaswamy, F. Beaufays, S. Augenstein, H. Eichner, C. Kiddon, D. Ramage, Federated learning for mobile keyboard prediction (2018). arXiv preprint arXiv:1811.03604

204. T.S. Brisimi, R. Chen, T. Mela, A. Olshevsky, I.C. Paschalidis, W. Shi, Federated learning of predictive models from federated electronic health records. Int. J. Med. Inf. **112**, 59–67 (2018)

205. L. Liu, C. Chen, Q. Pei, S. Maharjan, Y. Zhang, Vehicular edge computing and networking: a survey. Mobile Netw. Appl. **26**, 1145–1168 (2021)

206. D. Chen, X. Zhang, L.L. Wang, Z. Han, Prediction of cloud resources demand based on hierarchical pythagorean fuzzy deep neural network. IEEE Trans. Serv. Comput. (2019). https://doi.org/10.1109/TSC.2019.2906901

207. D. Chen, X. Zhang, L. Wang, Z. Han, Prediction of cloud resources demand based on fuzzy deep neural network, in *2018 IEEE Global Communications Conference (GLOBECOM)*, Abu Dhabi, UAE, 2018

208. Y. Pei, Y. Huang, Q. Zou, H. Zang, X. Zhang, S. Wang, Effects of image degradations to CNN-based image classification (2018). arXiv preprint arXiv:1810.05552

209. P. Dube, B. Bhattacharjee, S. Huo, P. Watson, B. Belgodere, J.R. Kender, Automatic labeling of data for transfer learning, in *IEEE Conference on Computer Vision and Pattern Recognition (CVPR) Workshops*, Long Beach, CA, 2019, pp. 122–129

210. J.A. Cortés-Osorio, J.B. Gómez-Mendoza, J.C. Riaño-Rojas, Velocity estimation from a single linear motion blurred image using discrete cosine transform. IEEE Trans. Instrum. Measur. **68**(10), 4038–4050 (2018)

211. T. Nishio, R. Yonetani, Client selection for federated learning with heterogeneous resources in mobile edge, in *ICC 2019–2019 IEEE International Conference on Communications (ICC)* (IEEE, Shanghai, 2019), pp. 1–7

212. N. Yoshida, T. Nishio, M. Morikura, K. Yamamoto, R. Yonetani, Hybrid-FL: cooperative learning mechanism using non-IID data in wireless networks (2019). arXiv preprint arXiv:1905.07210

213. S. Feng, D. Niyato, P. Wang, D.I. Kim, Y.-C. Liang, Joint service pricing and cooperative relay communication for federated learning, in *2019 International Conference on Internet of Things (iThings) and IEEE Green Computing and Communications (GreenCom) and IEEE Cyber, Physical and Social Computing (CPSCom) and IEEE Smart Data (SmartData)*, Atlanta, GA, 2019, pp. 815–820.

214. Y. Sarikaya, O. Ercetin, Motivating workers in federated learning: a stackelberg game perspective. IEEE Netw. Lett. **2**(1), 23–27 (2019)

215. Y. Zhang, L. Liu, Y. Gu, D. Niyato, M. Pan, Z. Han, Offloading in software defined network at edge with information asymmetry: a contract theoretical approach. J. Signal Process. Syst. **83**(2), 241–253 (2016)

216. Z. Hou, H. Chen, Y. Li, B. Vucetic, Incentive mechanism design for wireless energy harvesting-based internet of things. IEEE Internet Things J. **5**(4), 2620–2632 (2018)

217. S.R. Pandey, N.H. Tran, M. Bennis, Y.K. Tun, A. Manzoor, C.S. Hong, A crowdsourcing framework for on-device federated learning. IEEE Trans. Wirel. Commun. **19**(5), 3241–3256 (2020)

218. Y. Jing, B. Guo, Z. Wang, V.O. Li, J.C. Lam, Z. Yu, Crowdtracker: optimized urban moving object tracking using mobile crowd sensing. IEEE Internet Things J. **5**(5), 3452–3463 (2017)

219. D. Yang, K. Jiang, D. Zhao, C. Yu, Z. Cao, S. Xie, Z. Xiao, X. Jiao, S. Wang, K. Zhang, Intelligent and connected vehicles: current status and future perspectives. Sci. China Technol. Sci. **61**(10), 1446–1471 (2018)

220. J. Ren, G. Yu, G. Ding, Accelerating DNN training in wireless federated edge learning systems. IEEE J. Sel. Areas Commun. **39**(1), 219–232 (2020)

221. S. Dodge, L. Karam, Understanding how image quality affects deep neural networks, in *2016 Eighth International Conference on Quality of Multimedia Experience (QoMEX)* (IEEE, Lisbon, 2016), pp. 1–6

222. S. Ghosh, R. Shet, P. Amon, A. Hutter, A. Kaup, Robustness of deep convolutional neural networks for image degradations, in *2018 IEEE International Conference on Acoustics, Speech and Signal Processing (ICASSP)* (IEEE, Calgary, 2018), pp. 2916–2920

223. Y. Chen, S. He, F. Hou, Z. Shi, J. Chen, An efficient incentive mechanism for device-to-device multicast communication in cellular networks. IEEE Trans. Wirel. Commun. **17**(12), 7922–7935 (2018)

224. Y. Wang, M. Sheng, X. Wang, L. Wang, J. Li, Mobile-edge computing: partial computation offloading using dynamic voltage scaling. IEEE Trans. Commun. **64**(10), 4268–4282 (2016)

225. C. Li, S. Wang, X. Huang, X. Li, R. Yu, F. Zhao, Parked vehicular computing for energy-efficient internet of vehicles: a contract theoretic approach. IEEE Internet Things J. **6**(4), 6079–6088 (2019)

226. S. Wang, X. Huang, R. Yu, Y. Zhang, E. Hossain, Permissioned blockchain for efficient and secure resource sharing in vehicular edge computing (2019). arXiv preprint arXiv:1906.06319

227. T. Zhang, R.E. De Grande, A. Boukerche, Vehicular cloud: stochastic analysis of computing resources in a road segment, in *Proceedings of the 12th ACM Symposium on Performance Evaluation of Wireless Ad Hoc, Sensor, & Ubiquitous Networks*, New York, NY, 2015, pp. 9–16

228. A.F. Molisch, F. Tufvesson, J. Karedal, C.F. Mecklenbrauker, A survey on vehicle-to-vehicle propagation channels. IEEE Wirel. Commun. **16**(6), 12–22 (2009)

229. H.-Y. Lin, K.-J. Li, C.-H. Chang, Vehicle speed detection from a single motion blurred image. Image and Vis. Comput. **26**(10), 1327–1337 (2008)

230. Mnist handwritten digit database, 2010. [Online]. http://yann.lecun.com/exdb/mnist/

231. R. Timofte, K. Zimmermann, L. Van Gool, Multi-view traffic sign detection, recognition, and 3d localisation. Mach. Vis. Appl. **25**(3), 633–647 (2014)

232. J. Sun, W. Cao, Z. Xu, J. Ponce, Learning a convolutional neural network for non-uniform motion blur removal, in *Proceedings of the IEEE Conference on Computer Vision and Pattern Recognition*, Boston, 2015, pp. 769–777

233. I. Yaqoob, L.U. Khan, S.A. Kazmi, M. Imran, N. Guizani, C.S. Hong, Autonomous driving cars in smart cities: recent advances, requirements, and challenges. IEEE Netw. **34**(1), 174–181 (2019)

234. Federated learning, a step closer towards confidential AI. https://medium.com/frstvc/tagged/thoughts. Accessed 24 Jan 2020

235. L.U. Khan, Y.K. Tun, M. Alsenwi, M. Imran, Z.A, C.S. Hong, A dispersed federated learning framework for 6g-enabled autonomous driving cars (2021). arXiv preprint arXiv:2105.09641

236. A.K. Bairagi, N.H. Tran, W. Saad, Z. Han, C.S. Hong, A game-theoretic approach for fair coexistence between LTE-U and Wi-Fi systems. IEEE Trans. Veh. Technol. **68**(1), 442–455 (2019)

237. L. Zhou, On a conjecture by gale about one-sided matching problems. J. Econ. Theory **52**(1), 123–135 (1990)

238. A. Abdulkadiroğlu, T. Sönmez, Random serial dictatorship and the core from random endowments in house allocation problems. Econometrica **66**(3), 689–701 (1998)

239. S.A. Kazmi, N.H. Tran, W. Saad, Z. Han, T.M. Ho, T.Z. Oo, C.S. Hong, Mode selection and resource allocation in device-to-device communications: a matching game approach. IEEE Trans. Mobile Comput. **16**(11), 3126–3141 (2017)

240. G. Araniti, C. Campolo, M. Condoluci, A. Iera, A. Molinaro, LTE for vehicular networking: a survey. IEEE Commun. Mag. **51**(5), 148–157 (2013)

241. J. Wang, J. Liu, N. Kato, Networking and communications in autonomous driving: a survey. IEEE Commun. Surv. Tutorials **21**(2), 1243–1274 (2019)

242. S. Zhang, J. Chen, F. Lyu, N. Cheng, W. Shi, X. Shen, Vehicular communication networks in the automated driving era. IEEE Commun. Mag. **56**(9), 26–32 (2018)

243. M. Aledhari, R. Razzak, R.M. Parizi, F. Saeed, Federated learning: a survey on enabling technologies, protocols, and applications. IEEE Access **8**, 140699–140725 (2020)

244. L.U. Khan, M. Alsenwi, I. Yaqoob, M. Imran, Z. Han, C.S. Hong, Resource optimized federated learning-enabled cognitive internet of things for smart industries. IEEE Access **8**, 168854–168864 (2020)

245. M.N. Nguyen, N.H. Tran, M.A. Islam, C. Pham, S. Ren, C.S. Hong, Fair sharing of backup power supply in multi-operator wireless cellular towers. IEEE Trans. Wirel. Commun. **17**(3), 2080–2093 (2018)

246. N. World, https://www.networkworld.com/article/3407756/colocation-facilities-buck-the-cloud-data-center-trend.html. Accessed 20 Jan 2020

247. 3GPP, Evolved Universal Terrestrial Radio Access (E-UTRA): Physical Layer Procedures, Release 11. 3GPP standard TS 36.213, December 2012

248. Q. Wu, R. Zhang, Towards smart and reconfigurable environment: intelligent reflecting surface aided wireless network. IEEE Commun. Mag. **58**(1), 106–112 (2020)

249. C. Huang, A. Zappone, G.C. Alexandropoulos, M. Debbah, C. Yuen, Reconfigurable intelligent surfaces for energy efficiency in wireless communication. IEEE Trans. Wirel. Commun. **18**(8), 4157–4170 (2019)

250. C. Huang, S. Hu, G.C. Alexandropoulos, A. Zappone, C. Yuen, R. Zhang, M. Di Renzo, M. Debbah, Holographic mimo surfaces for 6g wireless networks: opportunities, challenges, and trends. IEEE Wirel. Commun. **27**(5), 118–125 (2020)

251. T.J. Cui, M.Q. Qi, X. Wan, J. Zhao, Q. Cheng, Coding metamaterials, digital metamaterials and programmable metamaterials. Light Sci. Appl. **3**(10), e218 (2014)

252. M. Cui, G. Zhang, R. Zhang, Secure wireless communication via intelligent reflecting surface. IEEE Wirel. Commun. Lett. **8**(5), 1410–1414 (2019)

253. Q. Nadeem, A. Kammoun, A. Chaaban, M. Debbah, M. Alouini, Large intelligent surface assisted MIMO communications (2019). arXiv preprint arXiv:1903.08127

254. A. Taha, M. Alrabeiah, A. Alkhateeb, Enabling large intelligent surfaces with compressive sensing and deep learning (2019). arXiv preprint arXiv:1904.10136
255. R.W. Heath, N. Gonzalez-Prelcic, S. Rangan, W. Roh, A.M. Sayeed, An overview of signal processing techniques for millimeter wave MIMO systems. IEEE J. Sel. Topics Signal Process. **10**(3), 436–453 (2016)

Printed in the United States
by Baker & Taylor Publisher Services